To David

A friend

Avigad

June - 1997

Spirulina platensis (Arthrospira)

Spirulina platensis (*Arthrospira*)

Physiology, cell-biology and biotechnology

Edited by

AVIGAD VONSHAK

Ben-Gurion University of the Negev, Israel

Taylor & Francis
Publishers since 1798

UK   Taylor & Francis Ltd, 1 Gunpowder Square, London EC4A 3DE
USA   Taylor & Francis Inc., 1900 Frost Road, Suite 101, Bristol, PA 19007

British Library Cataloguing in Publication Data

A catalogue record for this book is available from the British Library
ISBN 0-7484-0674-3 (cased)

Library of Congress Cataloging Publication Data are available

Cover design by Youngs Design in Production

Typeset in Times 10/12pt by MCS, Salisbury, Wiltshire

Printed in Great Britain by T. J. International Ltd

To my family: my parents Chaia and Nachman; my wife
Ahuva and boys Itai and Ohad for their trust,
support and love

Contents

Contents

Preface

Spirulina, or what was most likely *Arthrospira*, was rediscovered in the mid-1960s. A report by a botanist, Jean Léonard, who was a member of a French–Belgian expedition to Africa, described a blue-green cake sold in the food market of Fort Lamy, Chad. A further study revealed that this cake, called in the local dialect *dihé*, contained a blue-green alga identified as *Spirulina* and was consumed by the Kanembu tribe living along the alkaline lakes of Chad and Niger. Earlier reports in 1940 on the use of *dihé* had passed without any further notice.

At about the same time that Leonard had discovered *Spirulina* in Africa, a request was received from a company named Sosa Texcoco, by the Institut Français du Pétrole to study a bloom of algae occurring in the evaporation ponds of their sodium bicarbonate production facility in a lake near Mexico city. As a result, the first systematic and detailed study of the growth requirements and physiology of *Spirulina* was performed. This study, which was part of a Ph.D. thesis by Zarrouk, was the basis for establishing the first large-scale production plant of *Spirulina*. Nevertheless, most of the work never reached the attention of the international scientific community. If I may make personal comment, it is my impression that this was because the thesis was published in French and was never translated into English.

In a way, establishing the production site of *Spirulina* in Mexico was a kind of closing of a circuit. Some four hundred years before Sosa Texcoco had started the commercial production of *Spirulina*, in the sixteenth century, when the Spanish invaders conquered Mexico, they discovered that the Aztecs living in the Valley of Mexico in the capital Tenochtitlan were collecting a 'new food' from the lake. This food, described as a blue-green cake and called by the locals *tecuitlatl*, was used in a similar way to that for *dihé* of the Kanembu. Although no clear evidence is available on the exact composition of this food, it is clear that it contained a blue-green alga, and up to the present day *Spirulina maxima* is the dominant species in those waters.

So, from a semi-natural production facility in the early 1970s with a big variation in annual production from 100 to 400 tonnes, 25 years later *Spirulina* reached a total annual production exceeding 1000 tonnes per year with a forecast of doubling its market by the end of this century.

Preface

Developments in the field of algal biotechnology have yielded a few excellent contributions dealing with different aspects of the technology and some related problems in the mass cultivation of *Spirulina*. In the last ten years, a number of booklets have been published on *Spirulina*, most of them containing interesting information but mainly aimed at consumers, either written in a somewhat personal manner or containing practical information on the outdoor culturing of *Spirulina*.

There is no doubt that in order to support an industry producing 2000 tonnes of *Spirulina* biomass, with annual sales estimated at 40 million US$, much more has to be done, not only in the optimization of the production, but in more basic research in understanding the basics of the physiology, biochemistry and genetics of this alga.

This volume consists of two parts: the first deals with basic information on morphology, physiology, photosynthesis and genetics of laboratory cultures; the second part is dedicated to practical aspects of the biotechnology. It is the first time that a commercial production facility has revealed so many details on its production facility (Chapter 8). I can only hope that this will set an example to other producers to realize that at this stage of the game we do have to share information for the benefit of all interested in further development of this industry.

All the current mass producers of *Spirulina* are using open raceway ponds. It is my belief that with the increased demand for a high-quality product and a more sustainable and reliable production system, the future of mass production of *Spirulina*, as well as other algae, is in the development of closed systems. Which is the best one is still under investigation, and much more effort is required to further develop the optimal closed photobioreactor. This is the reason why two chapters (6 and 7) are devoted to this aspect. Both chapters were written by scientists from the Centro di Studio dei Microrganismi Autotrofi del CNR, Florence, Italy, which was headed by the late Prof. Florenzano, who was succeeded by Prof. Materrasi, pioneering the research in closed systems for outdoor cultivation of *Spirulina*.

This volume, which is a joint effort of many people involved in *Spirulina* research, is an attempt to collect the basic and the most relevant information on *Spirulina*. Some of it will be outdated by the time the book reaches the market, but we all hope that it will serve as a reference book and a starting point to further studies leading to a better understanding of algal biotechnology in general and mass production of *Spirulina* in particular. For this, I would like to give special thanks to all the contributors for their excellent work.

Last, I would like to thank two persons: Prof. Amos Richmond, a teacher and a friend, with whom I made my first steps some twenty years ago in algal biotechnology, and since then a source of inspiration and encouragement in all my studies. The second is Prof. Carl J. Soeder from Germany, with whose group I was first introduced to algal biotechnology. His pioneering work on outdoor production of *Scenedesmus* and *Chlorella* has benefitted many of us.

Bringing a collaborative volume to the printing stage is not only a job of writing and editing; it involves lots of administration, coordination and dedicated work. All of this and more has been contributed by Ms Ilana Saller, to whom I owe special thanks.

Avigad Vonshak

Sede-Boker, September 1996

Foreword

About one-third of world plant biomass consists of algae. The number of microalgae species is estimated at between 22 000 and 26 000, the biochemistry and ecophysiology of about fifty of which have been studied in detail. Microalgae represent all photosynthetic prokaryotic and eukaryotic microorganisms, and most of them live in aquatic environments.

Having used dense suspensions of the green unicellular alga, *Chlorella*, in experiments to study photosynthesis and having recognized that some microalgae could increase their biomass many times per day and that their dry matter could contain more than 50 per cent crude protein, tests were made in the early 1940s to grow microalgae on a large scale in Germany. Techniques for continuous cultivation were developed and attempts made to grow algae for potential commercial applications. In the early 1950s, researchers from the Carnegie Institution in Washington, DC, made outstanding contributions in this area, showing that the fat and protein contents of *Chlorella* cells could be modified by varying environmental conditions. Several pilot algal cultivation units were run at the Carnegie Institution. In 1957, Tamiya and his co-workers of the Tokugawa Institute of Biology, Tokyo, published their results on the outdoor mass cultivation of *Chlorella*, as part of an international project on algal culture also supported by the Carnegie Institution. In fact, Japan was the first country to produce and sell *Chlorella* biomass as a health food or as a water-soluble extract called 'Chlorella growth factor'.

In 1953, the German researchers of the Kohlenstoffbiologische Forschungstation e.b. (Essen, Germany) investigated the possibility of using the waste carbon dioxide produced by industrial plants of the Ruhr area to grow *Chlorella* sp. and another green unicellular alga, *Scenedesmus acutus*. Research was pursued in Dortmund, where Soeder and his team developed the technical know-how for mass cultivating algae, including a raceway with an endless channel in which the growth medium was stirred using a paddle wheel. In the early 1970s, the German Federal Government increased support to algal cultivation. Consequently, projects were launched in various developing countries, such as India, Peru and Thailand, where cultures of *Scenedesmus* relied on the technology developed in Dortmund.

In 1960, mass production of *Scenedesmus* was extended to Czechoslovakia, where a cascade-type cultivation apparatus was developed at the Institute of Hydrobotany, in Trěbǒn. A similar system was used in Bulgaria for mass producing *Chlorella*.

It was in the early 1960s that mass production of *Spirulina*, a filamentous blue-green alga or cyanobacterium, attracted the interest of researchers, with the pioneer work of Clement and his co-workers of the Institut Français du Pétrole (French Oil Institute). The French scientists were probably inspired by the observation that people living around Lake Chad, in the Kanem region, had been collecting *Spirulina maxima* for decades and using it in their diet.

In 1975, Oswald and his co-workers of the University of California, Berkeley, published their results on the simultaneous use of large-scale freshwater algal cultures for biomass (protein) production and wastewater treatment.

Microalgae can be used as sources of chemicals and biochemicals, such as polyunsaturated fatty acids, which are rare in plant and animal sources. In the food industry, there is a growing demand for various natural pigments, which can be produced from microalgae: in *Dunaliella* spp., beta-carotene may amount to as much as 14 per cent of its dry biomass; xanthophylls are particularly useful in pigmenting poultry, egg yolks and foodstuffs; several types of chlorophyll can find various uses; phycocyanin and phycoerythrin can be utilized as natural pigments in the food, drug and cosmetic industries to replace synthetic pigments. A food colorant containing phycocyanin from *Spirulina* is marketed by the Japanese company Dainippon Ink & Chemicals Inc. (DIC, Tokyo) under the brand name 'Lina blue-A'; the same company has also been producing another phycocyanin obtained from *Spirulina* but modified in order not to be water-soluble, so that it can be used for eye shadow, eye liner and lipstick, as it does not run when wet by water or sweat.

As far as polymers are concerned, heteropolysaccharides such as agar, alginic acid and carrageenan are extracted from red macroalgae on an industrial scale. But a red microalga, *Porphyridium aerugineum*, synthesizes and excretes a sulfonated polysaccharide at an output of up to 50 per cent of the rate of biomass production. A polysaccharide of this kind resembles carrageenan in its viscosity and emulsifying power; it is utilized for oil recovery from underground sand formations. Another polymer of potential use is poly-beta-hydroxybutyric acid (PHB), which occurs in a number of photosynthetic and non-photosynthetic bacteria. It is found in *Spirulina platensis*, where it accumulates during exponential growth to 6 per cent of the total dry weight.

It is therefore widely recognized that microalgae can be used as feed for animals and foodstuffs for humans and that their nutritional value, which has been substantiated by numerous studies, compares well with other conventional food products. In addition, there are many chemicals, biochemicals and pharmaceuticals which can be extracted from microalgae and which could be used in the food, pharmaceutical, cosmetics and chemical industries. Furthermore, the advantages offered by mass cultivation of microalgae over conventional crops have been highlighted: short cycle, continuous production throughout the year, easier genetic engineering and the possibility of growing microalgae in brackish water rather than freshwater, which is becoming a scarce resource.

What will determine the prospects of large-scale microalgaculture is not so much the market potential of its products, but their production cost. Efforts are therefore being made to lower the cost of the production unit by one order of magnitude or more, by increasing productivity through biotechnological research and innovation.

Current yields, which range from 10 tonnes to 30 tonnes of dry biomass per hectare per year and which are obtained in industrial-scale facilities, are low with regard to the theoretical maximum output and a reasonable optimum, which is a fraction of this maximum (over 200 tonnes of dry biomass per hectare per year).

The future of microalgaculture therefore depends, to a large extent, on two factors: first, the ability to reduce production costs and to make microalgal biomass a commodity traded in large quantities, but not limited to the health food market; second, the development of suitable bioreactors. Closed cultivation systems have several advantages over open raceways: cultures are better protected from contaminants, making the maintenance of monoalgal cultures easier; water loss and the subsequent increase in salinization of the culture medium are also avoided; areal volumes may be kept much smaller, due to greater cell densities, thereby reducing harvesting costs; optimal temperatures may be maintained more readily in closed systems, ensuring a greater output. Another requirement for successful large-scale cultivation of microalgae is the availability of a wider variety of microalgal species and of higher-performing strains responding favourably to varying environmental conditions existing outdoors. In the future, genetic engineering of microalgae could be a highly appreciated input to the success of microalgaculture.

Developing countries could benefit from current biotechnological improvements. Not only that but, with the expected drop in production costs, applications of microalgaculture could be extended beyond the present health food and fine biochemicals market to include, *inter alia*, an inexpensive high-protein supplement for human food. It is also likely that the increasing demand for special feedstuffs for animals (fish and aquatic organisms, poultry, swine and cattle) will contribute to further reducing production costs. This would also give a decisive impetus to mass microalgaculture in saline, highly saline and brackish waters unsuitable for the irrigation of most crop species and generally abundant in semi-arid regions. In the latter regions, the development of mass microalgaculture could offer a source of income and contribute to rational use of scarce natural resources.

Albert Sasson
Assistant Director-General,
Bureau of Studies, Programming and Evaluation,
UNESCO, Paris

Contributors

Dr Amha Belay
Earthrise Farms
P.O. Box 270
113 E. Hoober Road
Calipatria
92233 CA
USA

Zvi Cohen
Microalgal Biotechnology Laboratory
Jacob Blaustein Institute for Desert Research
Ben-Gurion University of the Negev
Sede-Boker Campus
84990 Israel

Joël de la Noüe
Faculté des Sciences de l'Agriculture et de l'Alimentation
Dept de Sciences et Technologie des Aliments
Université Laval
Cité Universitaire
Québec
Canada G1K 7P4

Kolli Bala Krishna
139 Uttarakhand
School of Life Sciences
Jawaharlal Nehru University
New Delhi 110 067
India

Contributors

Dr Gilles Laliberté
Faculté des Sciences de l'Agriculture et de l'Alimentation
Dept de Sciences et Technologie des Aliments
Université Laval
Cité Universitaire
Québec
Canada G1K 7P4

Dr Prasanna Mohanty
139 Uttarakhand
School of Life Sciences
Jawaharlal Nehru University
New Delhi 110 067
India

Eugenia J. Olguin
Department of Environmental Biotechnology
Institute of Ecology
A.P. 63, Xalapa
Veracruz
Mexico

Prof. A. Sasson
Assistant Director-General, BPE
UNESCO
7, Place de Fontenoy
75700 Paris
France

Madhulika Srivastava
139 Uttarakhand
School of Life Sciences
Jawaharlal Nehru University
New Delhi 110 067
India

Dr Luisa Tomaselli
Dept. di Scienze e Tecnologie
Alimentari e Microbioligiche
Centro di Studio dei Microrganismi Autotrofi
50144 Firenze
Piazzale delle Cascine 27
Italy

Dr Giuseppe Torzillo
Dept. di Scienze e Tecnologie
Alimentari e Microbioligiche
Centro di Studio dei Microrganismi Autotrofi
50144 Firenze
Piazzale delle Cascine 27
Italy

Dr Mario Tredici
Dept. di Scienze e Tecnologie
Alimentari e Microbioligiche
Centro di Studio dei Microrganismi Autotrofi
50144 Firenze
Piazzale delle Cascine 27
Italy

Ajay K. Vachhani
Microalgal Biotechnology Laboratory
Jacob Blaustein Institute for Desert Research
Ben-Gurion University of the Negev
Sede-Boker Campus
84990 Israel

Prof. Avigad Vonshak
Microalgal Biotechnology Laboratory
Jacob Blaustein Institute for Desert Research
Ben-Gurion University of the Negev
Sede-Boker Campus
84990 Israel

Graziella Chini Zittelli
Dept. di Scienze e Tecnologie
Alimentari e Microbioligiche
Centro di Studio dei Microrganismi Autotrofi
50144 Firenze
Piazzale delle Cascine 27
Italy

Morphology, Ultrastructure and Taxonomy of *Arthrospira (Spirulina) maxima* and *Arthrospira (Spirulina) platensis*

LUISA TOMASELLI

Introduction

Spirulina maxima Geitler and *S. platensis* Geitler are planktonic cyanobacteria that form massive populations in tropical and subtropical water bodies characterized by high levels of carbonate and bicarbonate and high pH (up to 11) (Busson, 1971; Iltis, 1970, 1971, 1980; Rich, 1931, 1933; Léonard and Compère, 1967; Marty and Busson, 1970; Clément, 1971, 1975; Durand-Chastel, 1980; Guérin-Dumartrait and Moyse, 1976). While *S. platensis* seems to be a more widely distributed species, mainly found in Africa, but also in Asia and South America, *S. maxima* (syn. *S. geitleri*) appears to be essentially confined to Central America. This latter species represents the main component of the phytoplankton of Lake Texcoco, which could be regarded as the original habitat of this species (Busson, 1971; Clément, 1975; Durand-Chastel, 1980; Guérin-Dumartrait and Moyse, 1976). Similarly, the alkaline saline lakes of the semidesert Sudan-Sahel zone, with epicenter in Lake Chad, and those of the Rift Valley, dominated by *S. platensis* water blooms, can be considered the starting points of this species (Iltis, 1970, 1971, 1980; Rich, 1931, 1933; Léonard and Compère, 1967; Marty and Busson, 1970; Clément, 1971, 1975; Guérin-Dumartrait and Moyse, 1976).

These two traditional species, currently included, according to Castenholz (1989), in the Genus *Arthrospira* Stizenberger 1852, Subsection III, Order *Oscillatoriales*, under the respective designation of *A. maxima* Setchell et Gardner 1917, and *A. platensis* (Nordst.) Gomont 1892, have maintained, since Geitler's revision of the *Cyanophyceae* (Geitler, 1925, 1932), the current incorrect designation of *Spirulina maxima* and *S. platensis*. This aspect will be treated more fully later on.

The continuing current designation of *Spirulina* for the species of the genus *Arthrospira* and particularly for its more important species (*A. platensis* and *A. maxima*) holds a more traditional, practical and technological meaning than a taxonomic one. This designation, however, often creates confusion. Therefore it is important to stress, considering the worldwide interest in this organism, that the genus *Arthrospira* is different and phylogenetically distant from *Spirulina*, although it shares with *Spirulina* the helical shape of the trichomes and the distribution of some species in the same selective habitats. The fame of this cyanobacterium is a result of its economic importance, which arises from the peculiar characteristics of the cultivated species (*A. platensis, A. maxima* or *A. geitleri* and/or *A. fusiformis*). These characteristics make the exploitation of this organism in mass culture very attractive as a source of food, feed, or fine chemicals (Florenzano, 1981; Jassby, 1988; Richmond, 1988; Seshadri and Jeeji Bai, 1992; Vonshak and Richmond, 1988) (see Chapters 5 and 8).

Even though the generic name *Arthrospira* is accepted, *ad interim*, as valid for the two species discussed here, *A. platensis* and *A. maxima*, they will frequently be referred to by the traditional names of *S. platensis* and *S. maxima* when reference is made to data from other authors.

Morphology

Arthrospira maxima Setchell et Gardner 1917 (syn. *S. maxima* Geitler 1932, *S. geitleri* De Toni 1935 or *Oscillatoria pseudoplatensis* Bourrelly 1970) and *A. platensis* (Nordst.) Gomont 1892 (syn. *S. platensis* Geitler 1925, *S. jenneri* var. *platensis* Nordstedt 1884 or *Oscillatoria platensis* Bourrelly 1970) are filamentous cyanobacteria recognizable by the main morphological feature of the genus: the arrangement of the multicellular cylindrical trichomes in an open left-hand helix along the entire length (Figures 1.1 and 1.2).

Under light microscopy, the blue-green non-heterocystous filaments, composed of vegetative cells that undergo binary fission in a single plane, show easily-visible transverse cross-walls. Filaments are solitary and free floating and display gliding motility. The trichomes, enveloped by a thin sheath, show more or less slightly pronounced constrictions at cross-walls and have apices either slightly or not at all attenuated. Apical cells may be broadly rounded or pointed and may be capitate and calyptrate. The width of the trichomes, composed of cylindrical shorter than broad cells, varies from about 6 to 12 μm (16 μm) in a variety of forms. The helix pitch (h) is determined by the equation

$$h = 2\,\pi\,r \cot \alpha \qquad\qquad (1.1)$$

where r is the radius of the cylinder surface to which the helix belongs and α is the angle formed by the helix and the cylinder generatrices and represents the slope of the helix curve. In many strains of these two species, the helix pitch varies from 12 to 72 μm. Also the helix diameter varies, ranging from about 30 to 70 μm (Pelosi et al., 1982).

Environmental factors, mainly temperature (Van Eykelenburg, 1979), physical and chemical conditions, may affect the helix geometry (Jeeji Bai and Seshadri, 1980; Jeeji Bai, 1985). One drastic alteration of this geometry is the reversible transition from helix to spiral shape, first observed by Van Eykelenburg and Fuchs (1980) after transferring the filaments from liquid to solid media. Although the helical shape of the trichome is considered a stable and constant property maintained in culture, there may be considerable variation in the degree of helicity between different strains of the same species and within the same strain. Even in natural monospecific populations, variations in trichome geometry may be observed.

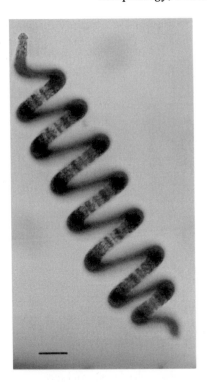

Figure 1.1 Light micrograph of *Arthrospira maxima*. Bar represents 20 μm.

Figure 1.2 Light micrograph of *Arthrospira platensis*. Bar represents 20 μm.

Moreover, straight or nearly straight spontaneous culture variants have been repeatedly reported (Jeeji Bai and Seshadri, 1980; Jeeji Bai, 1985; Lewin, 1986; Vonshak, 1986). Once a strain has converted to the straight form, both naturally or after physical or chemical treatments, such as UV radiation or chemicals (Pelosi et al., 1971), it does not revert back to the helical form. Jeeji Bai (1985) suggested that this is due to a mutation affecting some trichomes during certain growth conditions. The common occurrence of straight trichomes, in the cultures of *Arthrospira,* may suggest that the helicity character is carried on plasmids. However, no plasmids have been observed in the strains checked. When, in a culture of a helically coiled strain, a few filaments happen to become straight, they tend to become predominant. This is probably due to competition between the two morphones, as observed in *A. fusiformis* (Jeeji Bai, 1985).

As mentioned above, cell division occurs by binary fission on one plane at right angles to the long axis of the trichome. Trichome elongation occurs through multiple intercalary cell division all along the filament. Multiplication occurs only by fragmentation: the trichome breakage is transcellular by destruction of an intercalary cell, sacrificial cell (necridium), or lysing cell (Balloni et al., 1980; Tomaselli et al., 1981). Necridia act as unique specialized cells, allowing the transcellular breakup of a trichome, with formation of shorter segments or, occasionally, of hormogonia. Hormogonia cells undergo enlargement and maturation processes following a

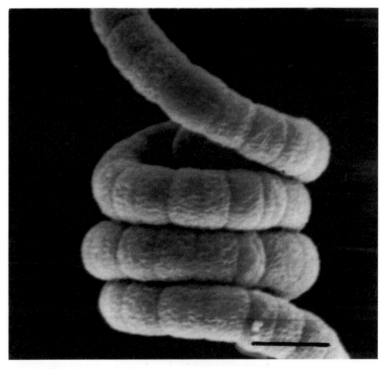

Figure 1.3 Scanning electron micrograph of *Arthrospira platensis* with part of trichome in continuous helical coil. Note the transverse cross-walls that divide the trichome into cells and the trichome surface covered by the thin sheath visible as shallow, anastomosed ribs. Bar represents 10 μm.

developmental cycle that was described by us (Balloni et al., 1980) and redrawn by Ciferri (1983).

The main distinctive features of the strains of *A. maxima* and *A. platensis* are here briefly described. In *A. maxima*, in contrast to *A. platensis*, the trichomes are not or very little constricted at the cross-walls; in addition, they have a wider diameter and are attenuated at the ends (Figures 1.1 and 1.2). In the mature trichome, the attenuation involves 6 or 7 cells; often the end cell is longer than broad, it is capitate and may be calyptrate for the thickening of the inner layer of the cell wall (Tomaselli et al., 1993). While in *A. platensis* the trichome architecture is usually characterized by a constant diameter of the loose coils, in *A. maxima* the coil diameter is slightly attenuated towards the apices. The free floating filaments of both *A. platensis* and *A. maxima* are densely granulated at the cross-walls because of the presence of gas vacuoles (aerotopes), but those of the latter species display a more regular disposition of this granulation. A thin sheath, invisible under light microscopy, envelops the trichomes. In the strains of *A. maxima* examined, the sheath seems to be thicker, so that under SEM analysis the trichome surface appears covered with deep ribs, whereas the trichome surface of *A. platensis* shows anastomosed, shallower ribs (Balloni et al., 1980) (Figure 1.3). *A. platensis* is characterized by short trichomes, usually with 5–7 coils. On solidified media, *A. maxima* displays higher gliding motility than *A. platensis*.

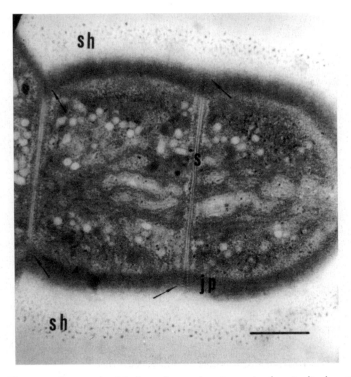

Figure 1.4 Electron micrograph of *Arthrospira maxima* in grazing longitudinal section along the trichome, illustrating the thin, diffluent, fibrillar, net-like sheath (**sh**) and the single row of junctional pores (**jp**) (arrows) flanking the cross-walls (**s**) around the circumference of the trichome. Bar represents 0.5 μm.

Ultrastructure

The cell organization of *A. maxima* and *A. platensis*, observed by electron microscopy, is typical of that of prokaryotic organisms, being devoid of a morphologically limited nucleus and of plastids and displaying an outer gram-negative type envelope, the cell wall (Marty and Busson, 1970; Van Eykelenburg, 1977, 1979; Tomaselli et al., 1993; Tomaselli et al., 1976).

Trichomes are surrounded by a thin, diffluent sheath. The sheath is about 0.5 μm thick and has a fibrillar, net-like structure (Figure 1.4). The sheath material, excreted through pores situated on the cell wall, has been thought to be involved in the filament motion (Van Eykelenburg, 1979).

The multilayered cell wall is thin, about 40–60 nm, and has an easily-detectable electron-dense layer corresponding to the peptidoglycan (Figure 1.5). The regularly spaced cross-walls that divide the trichome into cells, connected by plasmodesmata, are formed by centripetal ingrowth and extension of both the peptidoglycan and the more internal layer of the cell wall toward the centre of the cell. Thus, the cross-walls have a tripartite structure (Figures 1.4 and 1.6). Often, the formation of various cross-walls within the same cell can be observed. After treatment of the murein sacculus with sulphuric acid or in grazing longitudinal sections, junctional pores can be observed in the peptidoglycan layer next to the cross-wall (Figure 1.4). They are

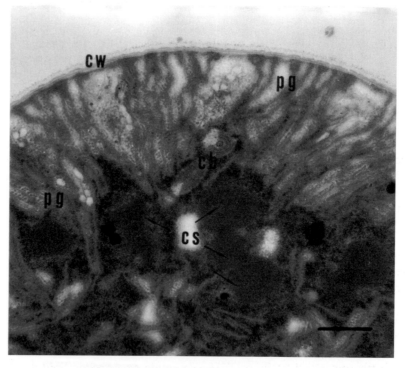

Figure 1.5 Electron micrograph of *Arthrospira platensis* in cross-section showing the multilayered cell wall (**cw**) and the subcellular organization of the cytoplasm. Note the accumulation of polyglucan granules (**pg**) close to the longitudinal cell wall (**cw**) and in the interthylakoidal space. In the central cytoplasmic region there are some carboxysomes (**cs**) (arrows) and two cylindrical bodies (**cb**). Bar represents 0.5 μm.

Figure 1.6 Electron micrograph of *Arthrospira maxima* in longitudinal section showing the division of the trichome into cells by cross-walls (**s**). Note the abundance of gas vacuoles (**gv**), the bundles of the thylakoid membranes (**t**), with associated phycobilisomes, many osmiophilic granules and ribosomes (**r**). Bar represents 0.5 µm.

arranged in one complete circular row, while in *Spirulina* the junctional pores are present in hemicircular rows in the concave region of the coil (Guglielmi and Cohen-Bazire, 1982a). Thus, pore distribution in the peptidoglycan layer constitutes a diacritical feature in distinguishing *Arthrospira* from *Spirulina* (Castenholz, 1989).

Just below the cell wall there is the plasma membrane, enclosing the cytoplasm, that is rich in subcellular inclusions typical of cyanobacteria (Carr and Whitton, 1973, 1982; Fogg et al., 1973; Jensen, 1984, 1985; Stanier (Cohen-Bazire), 1988; Stolz, 1991). These inclusions have a precise arrangement and distribution inside the cytoplasm (Tomaselli et al., 1993). The peripheral region of the cell is characterized by a low electron-dense cytoplasm mainly filled with polyglucan granules and gas vacuoles (Figures 1.5 and 1.6). There are also small osmiophilic granules, fibrils and lipid droplets. The thylakoid membranes, located between the peripheral and the central cytoplasm, are arranged in parallel and have associated electron-opaque phycobilisomes (Figure 1.6). The thylakoids, formed by two closely appressed unit membranes, appear as straight or sinuous bundles running parallel to the longitudinal wall and transversely to cross-walls. Low electron-dense thylakoid-free areas are filled with ribosomes and fibrils of DNA. Small electron-opaque lipid droplets are scattered among the thylakoid bundles and in the thylakoid-free areas (Figure 1.7). Sometimes, some spherical, highly osmiophilic, polyphosphate granules and large structured cyanophycin granules can be observed. The central electron-dense cytoplasmic region contains carboxysomes, recognizable by their polyhedral profile

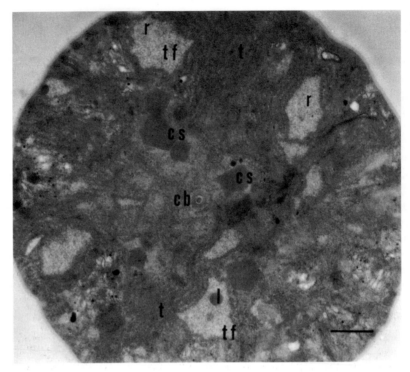

Figure 1.7 Electron micrograph of *Arthospira platensis* in cross-section showing the fine structure organization of the cytoplasm. Note the arrangement of the thylakoid membranes (**t**), the thylakoid-free areas (**tf**) filled with ribosomes (**r**) and lipid droplets (**l**), and the central nucleoplasmic region showing several carboxysomes (**cs**) and one cylindrical body (**cb**). Bar represents 0.5 μm.

and pseudocrystalline appearance, and some 'cylindrical bodies' (Pankratz and Bowen, 1963) (Figures 1.5 and 1.7). The cylindrical bodies appear to be sectioned both longitudinally and transversely. They are composed of a succession of alternating concentric electron-transparent shapeless layers and electron-dense structured layers. The structured layers are composed of double membranes. The cylindrical bodies, regularly found in *Arthrospira*, are reported as a characteristic of the *Oscillatoriaceae*, but they are not common in the cytoplasm of cyanobacteria (Van Eykelenburg, 1979; Jensen, 1984, 1985; Pankratz and Bowen, 1963; Chang and Allen, 1974).

Taxonomy

The common, incorrect designation of *Spirulina* for the genus *Arthrospira* and for its more widely studied species, discussed here, has been maintained since Geitler's revision (Geitler, 1925, 1932). Geitler unified within the priori genus *Spirulina* Turpin 1829 those oscillatoriacean organisms having the property of helically coiled trichomes along the entire length of the multicellular filaments, independent of the presence of more or less visible cross-walls under the light microscope. However, Stizenberger (1854) and Gomont (1892–1893) had previously placed the forms with regularly coiled filaments and visible septa within the genus *Arthrospira* Stizenberger 1852 and those with invisible septa within the genus *Spirulina* Turpin 1829. The difference in assessing the priority of the criteria used for classifying these organisms within the order *Oscillatoriales*, (Drouet, 1968; Hoffmann, 1985) led not only to the merging of *Arthrospira* into the genus *Spirulina* (Geitler, 1925, 1932; Hoffmann, 1985; Rippka et al., 1979) but also into *Oscillatoria* Vaucher 1904 along with *Spirulina* (Iltis, 1970, 1971; Bourrelly, 1970).

The view of *Arthrospira* and *Spirulina* as two separate genera has been shared in more recent times by several authors (Desikachary, 1959; Rippka et al., 1981; Anagnostidis and Komarek, 1988) and has been officially accepted by *Bergey's Manual of Systematic Bacteriology* (Castenholz, 1989). The separation between these two genera has been repeatedly affirmed, on the basis of many characteristics such as helicity and trichome size (Desikachary, 1959; Hindak, 1985), cell wall structure and pore pattern (Guglielmi and Cohen-Bazire, 1982a, 1982b), gas vesicles (Guglielmi et al., 1993), thylakoid pattern (Anagnostidis and Komarek, 1988), trichome motility and fragmentation (Anagnostidis and Komarek, 1988), GC content (Herdman et al., 1979) and oligonucleotide catalog of 16S rRNA (Guglielmi et al., 1993; Giovannoni et al., 1988; Waterbury, 1992). The complete sequence of the 16S rRNA gene and the internal transcribed spacer between the 16S and the 23S rRNA genes, determined for two *Arthrospira* strains (PCC 7345 and PCC 8005) and one *Spirulina* strain (PCC 6313) (Nelissen et al., 1994) and showing that two *Arthrospira* strains form a tight cluster not closely related to the *Spirulina* strain, clearly supports the separation of these two genera. To the above-mentioned diacritical characteristics, which are fully in agreement with our observations (Balloni et al., 1980; Tomaselli et al., 1976, 1993), should also be added the presence, solely in *Arthrospira,* of gamma-linolenic acid (Clément, 1975; Nichols and Wood, 1968; Paoletti et al., 1971; Materassi et al., 1980; Tredici et al., 1988; Cohen and Vonshak, 1991) and cylindrical bodies (Tomaselli et al., 1993).

The importance of recognizing *Arthrospira* as a quite different genus from *Spirulina* also arises from the need to avoid confusing species of this latter genus (e.g. *S. subsalsa*), only recently considered for nutritional purposes (Shimada et al., 1989), with those of *Arthrospira*, traditionally used for human consumption and whose safety has been ascertained (Chamorro Cevallos, 1980; Venkataraman and Becker, 1985).

The question of taxonomy and nomenclature of the most common *Arthrospira* species, *A. platensis*, *A. maxima* and/or *A. fusiformis*, was recently discussed by Komarek and Lund (1990) in the paper 'What is "*S. platensis*" in fact?'. In this work, to which reference should also be made for the exhaustive review and synonyms of the species in question, various isotype materials (algae exsiccatae), iconotypes and strains of *A. platensis* (Nordst.) Gomont 1892, *A. maxima* Setchell et Gardner 1917 and/or *S. fusiformis* Voronichin 1934 were revised. In this revision, Hindak's view (Hindak, 1985), that includes these organisms in two different taxonomic entities, is only partially confirmed: these are the benthic *Spirulina/ Arthrospira platensis* (species *A. platensis* Gomont) and the planktonic *Spirulina/ Arthrospira maxima, geitleri, fusiformis* (species *S. fusiformis* Voronichin). Komarek and Lund (1990) in agreement with Hindak (1985) maintain the inclusion of the non-planktonic-forms in the species *A. platensis* Gomont, but differentiate the planktonic forms on the basis of morphological characters and distribution into two different taxa: *A. maxima* Setchell et Gardner and *A. fusiformis* (Voronich.) comb.

nova (Komarek and Lund, 1990). However, it remains uncertain whether the planktonic taxa (*A. maxima* and *A. fusiformis*) can be considered to be different species.

While it is commonly agreed that wide morphological variations do occur in populations of *Arthrospira* spp., it is questioned whether considering *A. platensis* as a non-planktonic species, on the basis of the absence of aerotopes in the cells of dried isotypes, is fully justified.

The extensive studies carried out on *A. platensis* have never shown that this cyanobacterium, which forms massive water-blooms in several tropical lakes, could be considered non-planktonic (Busson, 1971; Iltis, 1970, 1971; Rich, 1931, 1933; Léonard and Compère, 1967; Marty and Busson, 1970; Clément, 1971). Therefore, it seems well-founded to continue to consider *A. platensis* as a planktonic species.

As Komarek and Lund (1990) report, the name *S. fusiformis* Voronichin 1934 was often used instead of *S. geitleri* De Toni 1935, since no significant differences were noted between these two species and Voronichin's name had priority. In fact, Hindak had included strains, classified as *S. maxima* or *S. geitleri* on the basis of the fusiform structure of the helices and the limited growth of trichome length, into the species *S. fusiformis* Voronichin (Hindak, 1985). Previously, Jeeji Bai and Seshadri (1980) had also included their fusiform isolate and the forms, earlier described as *A. platensis* by Rich (1931) and *A. platensis* f. *granulata* by Desikachary (1959), in *S. fusiformis* Voronichin. More recently, Desikachary and Jeeji Bai (1992), considering not only the fusiform shape of the helix but also the calyptrate condition of the filaments, proposed a new species, *A. indica*, for the calyptrate forms previously included in *S. fusiformis* Voronichin by Jeeji Bai and Seshadri. Therefore, at present, *A. indica* sp. nov. Desikachary and Jeeji Bai 1992, represents the latest proposed and described species that includes the fusiform calyptrate planktonic forms earlier described as *S./A. platensis* and *S./A. maxima, geitleri* or *fusiformis* (Desikachary and Jeeji Bai, 1992).

All this clearly highlights the confusion that exists concerning the species nomenclature and at the same time shows how much debate is still taking place over the classification of the traditional *Arthrospira* spp. Undoubtedly, the correct determination of species is a fundamental requisite for any type of research.

The 3rd edition of *Bergey's Manual* does not report, for the majority of cyanobacteria, any taxonomic key for the species of the genus *Arthrospira*. Only the two traditional species *A. maxima* Setchell et Gardner and *A. platensis* (Nordst.) Gomont are mentioned, along with the type species (Botanical Code) *A. jenneri* (Hass.) Stizenberger (Castenholz, 1989). This reflects the difficulty of species definition in the *Procaryotae*.

Applying the bacteriological approach to the classification of cultured cyanobacteria, which is one of the most diverse groups of prokaryotes, is very problematic. An improved classification, based not only on morphological and developmental features, has been hindered by many factors: the obligate photoautotrophy of many forms, the difficulty in growing a large variety of forms by using present culture methods, the scarce availability of pure culture, and so on.

The validity of the traditional morphological approach (Geitler's conception) (Anagnostidis and Komarek, 1985) is still recognized by Wilmotte and Golubic (Wilmotte and Golubic, 1991) for the relatively large and morphologically complex prokaryotes, such as the cyanobacterial species of *Arthrospira*. Maybe, as Hindak maintains (Hindak, 1985), the great morphological variability of the trichomes in the planktonic forms of this genus could in the future allow other features to be found for the differentiation of intraspecific taxa. However, in a classification based on morphological criteria, careful consideration must be given to the fact that many different cyanobacterial properties are the phenotype expression of the same genotype induced by environmental conditions. The same is also true for physiological properties of strains in pure culture. Therefore, although Anagnostidis and Komarek (1985) assert that the traditional botanical approach is the only way to identify the various taxa of cyanophytes, we should keep in mind the great advantages arising from the molecular techniques. In fact, the molecular approach, that allows us to understand the evolutionary relationships occurring among various microorganisms, represents the best way to develop an improved system of classification of the different living *Arthrospira* forms and pure cultures. Among the molecular methods using nucleic acids (DNA/DNA or DNA/RNA hybridization, DNA base composition, base sequence catalogs of oligonucleotides of 16S rRNA and 5S rRNA, ribopattern and DNA restriction fingerprint), the total DNA restriction fingerprint constitutes a powerful tool in analyzing strains belonging to the same genus. This method, recently successfully applied to the differentiation of other microorganisms and clustering at species level (Degli Innocenti et al., 1990; Giovannetti et al., 1990, 1992; Ventura et al., 1993) could greatly help in the definition of the taxonomic entities included in '*Spirulina platensis*' *sensu* Komarek and Lund (1990).

Nevertheless, the passage to the phylogenetic taxonomy is a slow process. Therefore, at present, taxonomists working on *Arthrospira* find themselves operating with two systems of classification: the traditional approach, still having a practical importance, and the molecular one, that allows the recognition of natural clusters.

Acknowledgements

The author is very much indebted to Prof. M. R. Palandri of the Dipartimento di Biologia Vegetale of the University of Florence for kindly providing the original TEM photographs and for her helpful discussion and interpretation. The author also wishes to express her thanks to Prof. R. Materassi, director of the Centro di Studio dei Microrganismi Autotrofi del CNR, and to her colleagues, for their critical comments on the manuscript. This research was supported by the National Research Council of Italy, Special Project RAISA, Sub-project No. 4, paper No. 1454.

References

ANAGNOSTIDIS, K. and KOMAREK, J. (1985) Modern approach to the classification system of cyanophytes 1. Introduction, *Arch. Hydrobiol.*, Suppl. 71, *Algol. Stud.*, **38/39**, 291.

ANAGNOSTIDIS, K. and KOMAREK, J. (1988) Modern approach to the classification system of cyanophytes 3, Oscillatoriales, *Arch. Hydrobiol.*, Suppl. 80, *Algol. Stud.*, **50/53**, 327.

BALLONI, W., TOMASELLI, L., GIOVANNETTI, L. and MARGHERI, M. C. (1980) Biologia fondamentale del genere *Spirulina*. In Materassi, R. (Ed.) *Prospettive della Coltura di Spirulina in Italia*, p. 49, Roma: CNR.

BOURRELLY, P. (1970) *Les Algues d'Eau Douce*, Vol. III: *Les Algues Bleues et Rouges, les Eugléniens, Péridiniens et Cryptomonadines*, p. 521, Paris: N. Boubée et Cie.

Spirulina platensis (Arthrospira)

BUSSON, F. (1971) Spirulina platensis (*Gom.*) *Geitler et* Spirulina geitleri *J. de Toni, Cyanophycées Alimentaires*, Service de Santé, Marseille.

CARR, N. G. and WHITTON, B. A. (1973) *The Biology of Blue-Green Algae*, Bot. Monogr. 9, Oxford: Blackwell Scientific.

CARR, N. G. and WHITTON, B. A. (1982) *The Biology of Cyanobacteria*, Bot. Monogr. 19, Oxford: Blackwell Scientific.

CASTENHOLZ, R. W. (1989) Subsection III, Order *Oscillatoriales*. In STANLEY, J. T., BRYANT, M. P., PFENNING, N. and HOLT, J. G. (Eds), *Bergey's Manual of Systematic Bacteriology*, Vol. 3, p. 1771, Baltimore: William and Wilkins.

CHAMORRO CEVALLOS, G. (1980) *Toxicological Studies on* Spirulina *alga: Sosa Texcoco, S.A., Pilot Plant for the Production of Protein from* Spirulina *Alga*, Rep. no. UF/MEX/78/048, UNIDO, Vienna.

CHANG, H. Y. Y. and ALLEN, M. M. (1974) The isolation of rapidosomes from the blue-green alga *Spirulina, J. Gen. Microbiol.*, **81**, 121.

CIFERRI, O. (1983) *Spirulina*, the edible micro-organism, *Microbiol. Rev.*, **47**, 557.

CLÉMENT, G. (1971) Une nouvelle algue alimentaire: la *Spirulina, Rev. Inst. Pasteur*, **4**, 103.

CLÉMENT, G. (1975) Production et constituants caractéristiques des algues *Spirulina platensis* et *maxima, Ann. Nutr. Alim.*, **29**, 477.

COHEN, Z. and VONSHAK, A. (1991) Fatty acid composition of *Spirulina* and *Spirulina*-like cyanobacteria in relation to their chemotaxonomy, *Phytochem.*, **30**, 205.

DEGLI INNOCENTI, F., FERDANI, E., PESENTI-BARILI, B., DANI, M., GIOVANNETTI, L. and VENTURA, S. (1990) Identification of microbial isolates by DNA fingerprinting: analysis of ATCC *Zymomonas* strains, *Biotechnol.*, **13**, 335.

DESIKACHARY, T. V. (1959) *Cyanophyta*, I.C.A.R., Monographs on Algae, p. 187, New Delhi.

DESIKACHARY, T. V. and JEEJI BAI, N. (1992) Taxonomic studies in *Spirulina*. In SESHRADI, C. V. and JEEJI BAI, N. (Eds), Spirulina *ETTA Nat. Symp.*, p. 12, Madras: MCRC.

DROUET, F. (1968) *Revision of the classification of the* Oscillatoriaceae, Monogr. 15, Acad. Nat. Sci, Philadelphia: Fulton Press.

DURAND-CHASTEL, H. (1980) Production and use of *Spirulina* in Mexico. In SHELEF, G. and SOEDER, C. J. (Eds), *Algae Biomass*, p. 39, Amsterdam: Elsevier/North-Holland Biomedical Press.

FLORENZANO, G. (1981) La coltura di *Spirulina*: antica fonte di proteine alimentari, *Riv. Oli, Grassi Derivati.*, **17**, 38.

FOGG, G. E., STEWART, W. D. P., FAY, P. and WALSBY, A. E. (1973) *The Blue-Green Algae*, London: Academic. Press.

GEITLER, L. (1925) Cyanophyceae. In PASCHER, A. (Ed.) *Süsswasserflora Deutschlands, Österreichs und der Schweiz*, 12, p. 342, Jena: FISCHER, G. Verlag.

GEITLER, L. (1932) Cyanophyceae. In RABENHORST, L. (Ed.) *Kryptogamenflora Deutschland, Österreich und der Schweitz*, 14, p. 916, Leipzig: Akad. Verlag.

GIOVANNETTI, L., FEDI, S., GORI, A., MONTAINI, P. and VENTURA, S. (1992) Identification of *Azospirillum* strains at the genome level with total DNA restriction pattern analysis, *System. Appl. Microbiol.*, **15**, 37.

GIOVANNETTI, L., VENTURA, S., BAZZICALUPO, M., FANI, R. and MATERASSI, R. (1990) DNA restriction fingerprint analysis of the soil bacterium *Azospirillum, Microbiol.*, **136**, 1161.

GIOVANNONI, S. J., TURNER, S., OLSEN, G. J., BARNS, S., LANE, D. J. and PACE, N. R. (1988) Evolutionary relationships among cyanobacteria and green chloroplasts, *J. Bacteriol.*, **170**, 3584.

GOMONT, M., 1892–1893, Monographie des *Oscillariées, Ann. Sci. Nat. Bot.*, **15**, 263, **16**, 91.

GUÉRIN-DUMARTRAIT, E. and MOYSE A. (1976) Charactéristiques biologiques des Spirulines, *Ann. Nutr. Alim.*, **30**, 489.

GUGLIELMI, G. and COHEN-BAZIRE, G. (1982a) Structure et distribution des pores et des perforations de l'enveloppe de peptidoglycane chez quelques cyanobactéries, *Protistologica*, **18**, 151.

GUGLIELMI, G. and COHEN-BAZIRE, G. (1982b) Etude comparée de la structure et de la distribution des filaments extracellulaires ou fimbriae chez quelques cyanobactéries, *Protistologica*, **18**, 167.

GUGLIELMI, G., RIPPKA, R. and TANDEAU DE MARSAC, N. (1993) Main properties that justify the different taxonomic position of *Spirulina* spp. and *Arthrospira* spp. among Cyanobacteria, *Bull. Inst. Oceanogr., Monaco*, **12**, 13.

HERDMAN, M., JANVIER, M., WATERBURY, J. B., RIPPKA, R., STANIER, R. Y. and MANDEL, M. (1979) Deoxyribonucleic acid base composition of cyanobacteria, *J. Gen. Microbiol.*, **111**, 63.

HINDAK, F. (1985) Morphology of trichomes in *Spirulina fusiformis* Voronichin from Lake Bogoria, Kenya, *Arch. Hydrobiol.*, Suppl. 71, *Algol. Stud.*, **38/39**, 201.

HOFFMANN, L. (1985) Quelques remarques sur la classification des *Oscillatoriaceae*, *Cryptogamie/Algologie*, **6**, 71.

ILTIS, A. (1970) Phytoplancton des eaux natronées du Kanem (Tchad). IV. Note sur les espèces du genre *Oscillatoria*, sous-genre *Spirulina* (Cyanophyta), *Cah. O.R.S.T.O.M., sér. Hydrobiol.*, **4**, 129.

ILTIS, A. (1971) Note sur *Oscillatoria* (sous-genre *Spirulina*) *platensis* (Nordst.) Bourrelly (Cyanophyta) au Tchad, *Cah. O.R.S.T.O.M., sér. Hydrobiol.*, **5**, 53.

ILTIS, A. (1980) Ecologie de *Spirulina platensis* dans les milieux natronés d'Afrique sahélienne. In Materassi, R. (Ed.) *Prospettive della Coltura di Spirulina in Italia*, p. 41, Roma: CNR.

JASSBY, A. (1988) *Spirulina*: a model for microalgae as human food. In LEMBI, C. A. and WAALAND, J. R. (Eds) *Algae and Human Affairs*, p. 149, Cambridge: Cambridge University Press.

JEEJI BAI, N. (1985) Competitive exclusion or morphological transformation? A case study with *Spirulina fusiformis*, *Arch. Hydrobiol.*, Suppl. 71, *Algol. Stud.*, **38/39**, 191.

JEEJI BAI, N. and SESHADRI, C. V. (1980) On coiling and uncoiling of trichomes in the genus *Spirulina*, *Arch. Hydrobiol.*, Suppl. 60, *Algol. Stud.*, **26**, 32.

JENSEN, T. E. (1984) Cyanobacterial cell inclusions of irregular occurrence: systematic and evolutionary implications, *Cytobios*, **39**, 35.

JENSEN, T. E. (1985) Cell inclusions in the Cyanobacteria, *Arch. Hydrobiol.*, Suppl. 71, *Algol. Stud.*, **38/39**, 33.

KOMAREK, J. and LUND, J. W. G. (1990) What is '*Spirulina platensis*' in fact?, *Arch. Hydrobiol.*, Suppl. 85, *Algol. Stud.*, **58**, 1.

LÉONARD, J. and COMPÉRE, P. (1967) *Spirulina platensis* (Gom.) Geitler, algue bleue de grande valeur alimentaire par sa richesse en protéines, *Bull. Jard. bot. Nat. Belg.*, **37**, 1.

LEWIN, R. A. (1980) Uncoiled variants of *Spirulina platensis* (Cyanophyceae: Oscillatoriaceae), *Arch. Hydrobiol.*, Suppl. 60, *Algol. Stud.*, **26**, 48.

MARTY, F. and BUSSON, F. (1970) Données cytologiques et systématiques sur *Spirulina platensis* (Gom.) Geitl. et *Spirulina Geitleri* J. De Toni (Cyanophyceae-Oscillatoriaceae), *C. R. Acad. Sc. Paris*, **270**, 786.

MATERASSI, R., PAOLETTI, C., BALLONI, W. and FLORENZANO, G. (1980) Some considerations on the production of lipid substances by microalgae and cyanobacteria. In SHELEF, G. and SOEDER, C. J. (Eds), *Algae Biomass*, p. 619, Amsterdam: Elsevier/North-Holland Biomedical Press.

NELISSEN, B., WILMOTTE, A., NEEFS, J-M. and DE WACHTER, R. (1994) Phylogenetic relationship among filamentous helical cyanobacteria investigated on the basis of 16S ribosomal RNA gene sequence analysis, *System. Appl. Microbiol.*, **17**, 206.

13

NICHOLS, B. W. and WOOD, B. T. B. (1968) The occurrence and biosynthesis of gamma-linolenic acid in a blue-green alga, *Spirulina platensis*, *Lipids*, **3**, 46.

PANKRATZ, H. S. and BOWEN, C. C. (1963) Cytology of blue-green algae. I. The cells of *Symploca muscorum*, *Am. J. Bot.*, **50**, 387.

PAOLETTI, C., MATERASSI, R. and PELOSI, E. (1971) Variazione della composizione lipidica di alcuni ceppi mutanti di *Spirulina platensis*, *Ann. Microbiol.*, **21**, 65.

PELOSI, E., MARGHERI, M. C. and TOMASELLI, L. (1982) Characteristics and significance of *Spirulina* morphology, *Caryologia*, **35**, 157.

PELOSI, E., PUSHPARAJ, B. and FLORENZANO, G. (1971) Mutazione di *Spirulina platensis* indotta dai raggi U.V. e da antibiotici, *Ann. Microbiol.*, **21**, 21.

RICH, F. (1931) Notes on *Arthrospira platensis*, *Rev. Algol.*, **6**, 75.

RICH, F. (1933) Scientific results of the Cambridge expedition to the East African Lakes (1930-1-7): the algae, *J. Limnol. Soc. Zool.*, **38**, 249.

RICHMOND, A. (1988) *Spirulina*. In BOROWITZKA, M. A. and BOROWITZKA, L. Y. (Eds), *Micro-algal Biotechnology*, p. 85, Cambridge: Cambridge University Press.

RIPPKA, R., DERUELLES, J., WATERBURY, J. B., HERDMAN, M. and STANIER, R. Y. (1979) Generic assignments, strain histories and properties of pure culture of cyanobacteria, *J. Gen. Microbiol.*, **111**, 1.

RIPPKA, R., WATERBURY, J. B. and STANIER, R. Y. (1981) Provisional generic assignments for Cyanobacteria in pure culture. In STARR, M. P., STOLP, H., TRUPER, H. G., BALOWS, A., and SCHLEGEL, H. G. (Eds), *The Prokaryotes*, Vol. 1, p. 247, Berlin: Springer-Verlag.

SESHADRI, C. V. and JEEJI BAI, N. (1992) *Spirulina*, *ETTA Nat. Symp.*, Madras: MCRC.

SHIMADA, A., OGUCHI, M., OTSUBO, K., NITTA, K., KOYANO, T. and MIKI, K. (1989) Application of tubular photo-bioreactor system to culture *Spirulina* for food production and gas exchange. In MIYACHI, S., KARUBE, I. and ISHIDA, Y. (Eds), *Current Topics in Marine Biotechnology*, p. 147, Tokyo: Jap. Soc. Mar. Biotechnol.

STANIER (COHEN-BAZIRE), G. (1988) Fine structure of Cyanobacteria, *Meth. Enzymol.*, **167**, 157.

STIZENBERGER, E. (1854) *Spirulina* und *Arthrospira* (nov. gen.), *Hedwigia*, **1**, 32.

STOLZ, J. F. (1991) *Structure of Phototrophic Prokaryotes*, Florida: CRC Press.

TOMASELLI FEROCI, L. and BALLONI, W. (1976) *Spirulina labyrinthiformis* Gomont: prima segnalazione nelle Terme di Segesta (Sicilia), *Gior. Bot. It.*, **110**, 241.

TOMASELLI FEROCI, L., MARGHERI, M. C. and PELOSI, E. (1976) Die Ultrastruktur von *Spirulina* im Vergleich zu *Oscillatoria*, *Zbl. Bakt.* Abt. II, **131**, 592.

TOMASELLI, L., GIOVANNETTI, L. and MARGHERI, M. C. (1981) On the mechanism of trichome breakage in *Spirulina platensis* and *S. maxima*, *Ann. Microbiol.*, **31**, 27.

TOMASELLI, L., PALANDRI, M. R. and TANI, G. (1993) Advances in preparative techniques for observation of the fine structure of *Arthrospira maxima* Setch. et Gardner (syn. *Spirulina maxima* Geitler), *Arch. Hydrobiol.*, Suppl. 100, *Algol. Stud.*, **71**, 43.

TREDICI, M. R., MARGHERI, M. C., DE PHILIPPIS, R., BOCCI, F. and MATERASSI, R. (1988) Marine cyanobacteria as a potential source of biomass and chemicals, *Int. J. Solar Energy*, **6**, 235.

VAN EYKELENBURG, C. (1977) On the morphology and ultrastructure of the cell wall of *Spirulina platensis*, *A. Leeuwenhoek*, **43**, 89.

VAN EYKELENBURG, C. (1979) The ultrastructure of *Spirulina platensis* in relation to temperature and light intensity, *A. Leeuwenhoek*, **45**, 369.

VAN EYKELENBURG, C. and FUCHS, A. (1980) Rapid reversible macromorphological changes in *Spirulina platensis*, *Naturwissenschaften*, **67**, 200.

VENKATARAMAN, L. V. and BECKER, E. W. (1985) Toxicological studies. In *Biotechnology and Utilization of Algae, The Indian Experience*, p. 186, New Delhi: Dept. Sci. Technol.

VENTURA, S., GIOVANNETTI, L., GORI, A., VITI, C. and MATERASSI, R. (1993) Total DNA restriction pattern and quinone composition of members of the family *Ectothiorhodospiraceae, System Appl. Microbiol.*, **16**, 405.

VONSHAK, A., personal communication (1986).

VONSHAK, A. and RICHMOND, A. (1988) Mass production of *Spirulina* – an overview. *Biomass*, **15**, 233.

WATERBURY, J. B. (1992) The cyanobacteria – isolation, purification and identification. In BALOWS, A., TRUPER, H. G., DWORKIN, M., HARDER, W. and SCHLEIFER, K. H. (Eds), *The Prokaryotes*, Vol. II, p. 2058, New York: Springer-Verlag.

WILMOTTE, A. and GOLUBIC, S. (1991) Morphological and genetic criteria in the taxonomy of Cyanophyta/Cyanobacteria, *Arch. Hydrobiol.*, Suppl. 92, *Algol. Stud.*, **64**, 1.

2

The Photosynthetic Apparatus of *Spirulina:* Electron Transport and Energy Transfer

PRASANNA MOHANTY, MADHULIKA SRIVASTAVA AND KOLLI BALA KRISHNA

Introduction

Photosynthesis is a metabolic process by which all photoautotrophic organisms, i.e. photosynthetic bacteria (*Rhodopseudomonas* sp., *Rhodospirillum* sp., etc.), cyanobacteria (*Spirulina* sp., *Synechococcus* sp., *Anabaena* sp., etc.) and higher plants, are able to convert light energy into chemical energy in the form of carbohydrates. During oxygenic photosynthesis in cyanobacteria and higher plants, light energy is utilized to transport electrons from water to $NADP^+$ with a concomitant evolution of oxygen. The ATP and NADPH generated during this light-driven process are subsequently used for enzymatic conversion of atmospheric CO_2 to carbohydrates. Thus, the overall reaction carried out during photosynthesis may be considered to be

$$6CO_2 + 12H_2O \xrightarrow{\text{light energy}} C_6H_{12}O_6 + 6O_2 + 6H_2O \qquad (2.1)$$

The site of the light reactions of photosynthesis is the thylakoid membranes.

Light Reactions and the Photosynthetic Apparatus

Light Harvest and Energy Transfer

The light energy utilized during photosynthesis is captured by photosynthetic pigments such as chlorophyll, phycocyanobilin, carotenoids, etc. All these pigments have an extended array of conjugated bonds. This allows them to interact with and efficiently absorb electromagnetic radiation in the visible range. The spectral qualities of these pigments are affected not only by their chemical composition, but also by the protein milieu around them. Typically, the absorption bands of the pigments complexed with the proteins are broadened because of further splitting of energy levels. This leads to a more efficient harvesting of the light energy.

While chlorophyll and carotenoids are complexed with proteins that are an integral part of the photosystem I and photosystem II in cyanobacteria, a specialized supramolecular pigment protein complex called the phycobilisome (PBsome) (Glazer, 1982, 1984, 1985, 1989; Gantt, 1988; Bryant, 1991) is also present. It binds phycocyanobilin and other similar chromophores (bilins). Because of the presence of bilins as chromophores, these antenna systems are able to harvest the energy from light regimes where Chl *a* does not absorb or absorbs only very weakly.

The bilin chromophores (phycocyanobilin, phycoerythrobilin and others) have an open-chain tetrapyrrole structure and are linked covalently (by a thioether bond) to the cystein residues of the apoprotein. Spectral characteristics of the various biliproteins (Table 2.1) are widely different owing to the differences in:

1   the chemical structure of the chromophore;

2   its specific association with the apoprotein; and

3   interactions with other biliproteins and linker polypeptides.

As a result, PBsomes are capable of absorbing light energy from nearly 550 nm to 650 nm. Another important feature is the directed transfer of this absorbed energy to the reaction center (RC). PBsomes seem to be extremely efficient in this respect because of a very ordered structure. The biliproteins absorbing higher energy radiation are spatially most distal to the RC, while those which absorb a lower energy radiation are most proximal. Also, the biliproteins are arranged in arrays such that the emission spectrum of one kind has a large overlap with the absorption spectrum of the other biliprotein, thus facilitating energy transfer towards the RC.

The basic subunit of the apoprotein is a heterodimer of polypeptides α and β. The $\alpha\beta$ unit has a tendency to form trimers and hexamers $(\alpha\beta)_6$. Many hexamers are linked by linker polypeptides to form rods, which are arranged around a core comprising allophycocyanin hexamers. Figure 2.1 shows the polypeptide pattern of purified

Table 2.1   Properties of major cyanobacterial phycobiliproteins (adapted from Bryant, 1991)

Protein	Subunit designation[a]	Molecular mass (kDa)[b]	Absorption maximum (nm)[b]	Fluorescence maximum (nm)[b]
Core components				
Allophycocyanin	a^{AP} β^{AP}	100	650	660
Allophycocyanin β	a^{AP-B} β^{AP}	89	670–618	675
Allophycocyanin β^{18}	a^{AP} β^{18}	35–70	616	640
Core linker phycobiliprotein	L_{CM}	72–120	665	680
Peripheral rod components				
Phycocyanin	a^{PC}	36.5–220	620	640

[a] Subunit designations follow the recommendations of Glazer (1984), a and β denote the subunits of the protein. L and CM stand for linker polypeptide and core-membrane junction, respectively. [18]is AP-like polypeptide with an apparent mass of about 18 kDa.
[b] These values are for the aggregation states.

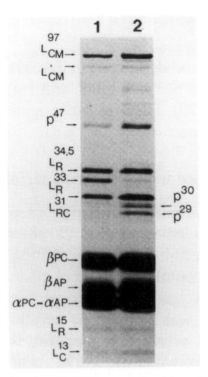

Figure 2.1 Polypeptide profile of purified PBsomes isolated from *Spirulina maxima* cells grown under 60 μE m^{-2}s^{-1} (lane 1) and 400 μE m^{-2}s^{-1} (lane 2). Polypeptides were separated in LDS-PAGE and stained with coomassie blue. L, linker; R, rod; CM, core-membrane; RC, rod-core; PC, phycocyanin; AP, allophycocyanin; α and β, subunits of heterodimer; superscript numbers refer to molecular weights in kDa (reproduced with permission from Garnier et al., 1994).

PBsomes in LDS-PAGE. The core is most proximal to the reaction center and also has an emission spectrum that overlaps well with the Chl *a* absorption spectrum (Mohanty et al. 1985). Thus, the core, along with a linker polypeptide, is responsible for the final step of energy transfer from PBsomes to the photosystem II (PS II).

One of the important features of cyanobacterial PBsomes is that their size and composition may be altered depending on the growth conditions, especially the spectral quality of the irradiating light (Hattori and Fujita, 1959; Bennett and Bogorad, 1973; Tandeau de Marsac, 1991). Nutritional status of the growth medium also affects to a great extent the size and composition of the PBsomes (Grossman et al., 1993).

An individual chlorophyll molecule absorbs only two or three photons per second even under direct solar illumination. No living organism could grow if their RC did not have light-harvesting antenna molecules in excess. Emerson and Arnold (1932) have demonstrated that hundreds of chlorophyll molecules cooperate to harvest solar energy during photosynthesis. However, the ability to undergo charge separation is endowed on certain specialized chlorophyll molecules present in the RC. Hence, it is important for the energy absorbed by each of the chlorophyll molecules to reach the RC. Duysens (1952) demonstrated that the absorbed energy migrates between any two chromophores to a considerable extent via the lowest excited singlet state. In the case of chlorophylls, optimal conditions prevail for the Forster type of energy transfer, and it is believed to be the primary mechanism by which excitation energy migrates from the antenna to the RC (van Grondelle and Amesz, 1986). For two molecules about 2 nm apart, the rate of energy transfer, K is about 5×10^{11} s^{-1}. In general, where r is the distance between interacting molecules,

$$K \propto r^{-6} \tag{2.2}$$

The efficiency of utilization of the absorbed radiation by photochemistry has been suggested to be more than 90 per cent (Bjorkman and Demming, 1987). Two other major routes of deexcitation that compete with photochemistry are fluorescence and thermal deactivation. While the quantum yield of fluorescence from Chl *a* in ether is about 30 per cent, the maximum yield in the photosynthetic apparatus is only about 3 per cent when the reaction centers are closed. The value is only 0.6 per cent when the photochemistry is fully operative. Such a high efficiency of utilization of the absorbed energy was achieved in the reaction centers on account of the protein environment in the RC. The quantum yield of photochemistry ($\emptyset_P$) is given by,

$$\emptyset_P = \frac{K_P}{K_P + K_D + K_F} \tag{2.3}$$

where K_P is the rate constant for photochemistry, K_D is the rate constant for thermal relaxation, and K_F is the rate constant for fluorescence.

The fact that fluorescence and photochemistry compete with each other for deexcitation makes the former a very useful tool to probe and evaluate the photochemical process under various conditions.

Electron Transport

The process of photosynthesis comprises two distinct phases. The light reactions are involved in capture of light energy and its conversion to ATP and reducing power. This is accomplished by the light-mediated transport of electrons from H_2O to NADP$^+$. The site for the reactions involved in the electron transport is the thylakoid membranes dispersed in the cell. Unlike higher plants, these membranes are not organized into stacked (grana) and unstacked (stroma) regions in the cyanobacteria. The dark reactions are purely enzymatic in nature and utilize the energy generated by the light reactions to convert CO_2 to carbohydrates. These reactions are carried out in the stroma of the chloroplast (in higher plants) and in the cytoplasm, in the case of cyanobacteria.

The light is absorbed by the pigments present in the antennae of photosystem I and II (PS I and PS II). The excitation energy transferred to the reaction center of the photosystems initiates charge separation. Secondary electron flow ensues through various reducible components of the photosynthetic electron transport chain. Essentially, these components are able to mediate the flow of electrons from water to NADP$^+$. Concomitantly, transthylakoid electrochemical potential difference for protons is created which is eventually utilized for phosphorylation of ADP. The ATP and NADPH thus generated are used for CO_2 fixation.

The transport of electrons from water to NADP$^+$ is mediated by a number of proteins, pigment–protein complexes and other organic molecules which function in a highly coordinated manner. The complex set of reactions has been described in the Z scheme of electron transport (Figure 2.2) proposed initially by Hill and Bendall (1960). The hallmarks of the Z scheme are:

- utilization of electromagnetic energy at two sites, i.e. PS I and PS II;
- transport of electrons in a non-cyclic manner from H_2O to $NADP^+$; and
- generation of ATP and NADPH.

In addition to the non-cyclic electron transport, the photosynthetic apparatus of *Spirulina* sp. and several other cyanobacteria is also known to carry out light-mediated cyclic transport of electrons around PS I. Such electron transport generates a transthylakoid pH gradient (and subsequently ATP), but is unable to generate reducing power in the form of NADPH. Thylakoid membranes of *Spirulina* and other cyanobacteria are also the site for respiratory activity; the two electron transport systems interact closely as some of the components are common between them (Scherer, 1990).

Intact trichomes of *Spirulina* grown in alkaline medium either in low photon flux density (PFD) or in high PFD can support endogenous O_2 evolution upon illumination (Table 2.2). Photosynthetically active cultures support PS II-dependent O_2 evolution in the presence of a variety of oxidants, including several derivatives of benzoquinone, like *p*-benzoquinone, phenylbenzoquinone, etc. Although these cells yield high rates of O_2 evolution, no serious attempts have been made to measure action spectra or quantum yield of the benzoquinone-supported Hill reaction. The *Spirulina platensis* trichomes can also support the whole chain electron flow from H_2O to the PS I terminal acceptor, MV, at reasonable rates. Although MV does not easily enter into cytosolic space in many cyanobacteria, this acceptor supports light-dependent O_2 uptake in *Spirulina* sp., possibly due to the close proximity of thylakoid membranes to the plasma membrane.

A variety of preparative methods (Shyam and Sane, 1989; Murthy, 1991; Vonshak et al., 1988) are available for spheroplast and thylakoid preparation of *Spirulina* sp. *Spirulina platensis* spheroplasts have been used to monitor PS II, whole chain and PS I-catalyzed photochemical reactions (Table 2.2). Phycobilin-free thylakoid

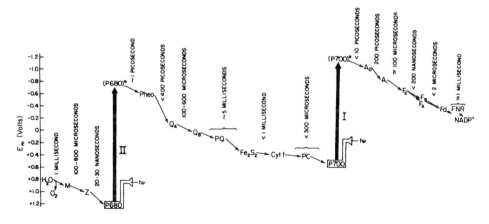

Figure 2.2 An updated version of the Z-scheme of non-cyclic photosynthetic electron transport showing the lifetimes of various electron transfer reactions (reproduced with permission from Govindjee and Eaton-Rye, 1986).

Table 2.2 Typical light-saturated rates of photosynthetic function as measured by O_2 polarography in *Spirulina*

Species and/or preparation	Reaction[a]	Light-saturated rates (O_2)	Reference
A. Intact cells			
1. *S. platensis*	endogenous O_2 evolution coupled to CO_2 fixation	3.5 µmol mg dry wt^{-1} h^{-1} (low light grown cells, 50 µmol m^{-2} s^{-1}, LLG) and 5 µmol mg dry wt^{-1} h^{-1} (high light grown cells, HLG)	Shyam and Sane, 1989
2. a. *S. platensis*	$H_2O \rightarrow MV$	285 ± 15 µmol mg chl^{-1} h^{-1}	Murthy and Mohanty, 1995
b. *S. platensis*	$H_2O \rightarrow pBQ$	405 ± 21 µmol mg chl^{-1} h^{-1}	Murthy and Mohanty, 1995
B. Spheroplasts			
1. *S. platensis*	PS II activity $H_2O \rightarrow FeCN$ (DBMIB)	1200 ± 85 µmol mg dry wt^{-1} h^{-1} (in LLG) and 2300 ± 58 µmol mg dry wt^{-1} h^{-1} (in HLG)	Shyam and Sane, 1989
2. *S. platensis*	PS I rate $DCPIPH_2 \rightarrow MV$	1400 ± 40 µmol mg dry wt^{-1} h^{-1} (in LLG) and 2000 ± 38 µmol mg dry wt^{-1} h^{-1} (in HLG)	Shyam and Sane, 1989
C. Thylakoid preparations			
1. *S. platensis*	PS II assay $H_2O \rightarrow pBQ$	260 µmol mg chl^{-1} h^{-1}	Murthy and Mohanty, 1995
2. *S. maxima*	PS I $TMPD/ASC \rightarrow MV$ $DAD/ASC \rightarrow MV$ $DCPIP/ASC \rightarrow MV$	1193 ± 3 µmol mg chl^{-1} h^{-1} 1018 ± 45 µmol mg chl^{-1} h^{-1} 811 ± 33 µmol mg chl^{-1} h^{-1}	Murthy and Mohanty, 1995; Murthy, 1991
3. *S. maxima*	$H_2O \rightarrow MV$[b] $H_2O \rightarrow FeCN$	75 µmol mg chl^{-1} h^{-1} 55 µmol mg chl^{-1} h^{-1}	Lerma and Gomez-Lojero, 1982
4. *S. maxima*	Photophosphorylation (cyclic) Photophosphorylation (cyclic, initial rate)	921 µmol ATP mg chl^{-1} h^{-1} 72–168 µmol ATP mg chl^{-1}h^{-1}	Lerma and Gomez-Lojero, 1982; Bakels et al., 1993

[a] MV, Methyl viologen; pBQ, *p*-benzoquinone; FeCN, $K_3[Fe(CN)_6]$; DBMIB, 2,5-dibromo-3-methyl-6-isopropyl-*p*-benzoquinone; DCPIP, 2,6-dicholorophenol indophenol; $DCPIPH_2$, reduced 2,6-dichlorophenol indophenol; TMPD, *N,N,N',N'*-tetramethyl phenylene diamine; ASC, ascorbic acid; DAD, diaminodurene; DCPIP, 2,6-dichlorophenol indophenol.
[b] A high dark O_2 uptake was observed in stored thylakoids.

membranes can easily be prepared by using French Press, and these thylakoid membranes support PS II-catalyzed O_2 evolution and PS I-catalyzed MV reduction and $NADP^+$reduction.

Organization and Function of Supramolecular Complexes

The various supramolecular complexes involved in the electron transport in cyanobacterial thylakoids are: PS II, PS I, cytochrome b_6/f complex, ATPase and the PBsomes (light-harvesting pigment–protein complexes). In addition to these, two mobile electron carriers are also present, namely, plastoquinone (PQ) and plastocyanin.

Photosystem II

This multisubunit–pigment protein complex acts as the water-plastoquinol-oxidoreductase catalyzing the reaction

$$4H^+_{(s)} + 2PQ + 2H_2O + 4h\nu \rightarrow 2PQH_2 + O_2 + 4H^+_{(l)} \tag{2.4}$$

Despite a considerable evolutionary distance between cyanobacteria and higher plants, the PS II complex of the two groups is very similar. More than 22 polypeptides are known to constitute the PS II (Hansson and Wydrzynski, 1990; Vermaas and Ikeuchi, 1991). Functional roles have been assigned to some of them while others are presently thought to be important only from a structural point of view. The RC from purple bacteria has been crystallized and its detailed structure is known (Deisenhoefer and Michel, 1989). This has proven to be a very useful model for PS II. The analogy between these two reaction centers has been demonstrated using site-directed mutagenesis (Debus et al., 1988a, 1988b; Vermaas et al., 1988b; Rutherford and Zimmerman, 1984) and spectroscopic techniques (Rutherford and Zimmerman, 1984; Schatz and van Gorkom, 1985). Based on function, the organization of PS II may be considered as follows.

Reaction center The minimum unit capable of a stable photo-induced charge separation has been isolated by Nanba and Satoh (1987) from spinach and by Barber et al. (1987) from pea. *Synechocystis* 6803 was the first cyanobacterium from which the RC was isolated (Gounaris et al., 1989). It comprises a heterodimer of closely related polypeptides D1 and D2 in addition to one cytochrome b_{559} heme iron, 8 Chl *a* molecules and one pheophytin *a* molecule. The two polypeptides D1 and D2 provide the protein environment within their core, such that P680 (a specialized chlorophyll dimer) is able to undergo photooxidation. D1 and D2 also bind other cofactors like pheophytin, Q_A, Q_B, etc., in appropriate orientation to effect stabilization of the charge-separated state. Many of the known herbicides, such as DCMU, atrazine, etc., interfere with Q_B binding by interacting with the D1 polypeptide (Tischer and Strotmann, 1977; Trebst, 1986).

Regulatory cap The unique feature of PS II as a RC is its ability to evolve O_2 by oxidizing H_2O. In cyanobacteria, a 33 kDa polypeptide is involved in stabilizing the manganese atoms necessary for the splitting of H_2O and is thus called the manganese-stabilizing polypeptide (MSP). In addition to MSP, higher plants have

18 kDa and 24 kDa polypeptides as part of the water-oxidizing complex located on the luminal side of PS II. Since these polypeptides also play a role in regulating the ionic requirements for O_2 evolution, they have been named the regulatory cap (Hansson and Wydrzynski, 1990).

Proximal antenna Two polypeptides, called CP47 and CP43, constitute the proximal antenna for the PS II RC. The exact number of Chl *a* molecules bound by these polypeptides has not yet been clearly established. The estimates of Chl *a* molecules per CP47 polypeptide range from 6–11 (Tang and Satoh, 1984) to 20–30 (Yamaguchi et al., 1988). For CP43, the number of Chl *a* molecules has been reported to be 11 (Akabori et al., 1988) and 26 (Yamaguchi et al., 1988). These Chl *a* molecules harvest light energy and transfer it to the RC. They are also supposed to be involved in the energy transfer from light harvesting complex (PBsomes or LHCP II) to the RC. Recent studies suggest the role of CP43 and CP47 in the assembly of PS II (Vermaas et al., 1988a). Cyanobacterial mutants in which CP43 has been impaired due to mutation of *psb C* gene, cannot grow photoautotrophically. However, the thylakoids from such mutants of *Synechocystis* 6803 do support light-induced reduction of DCIP in presence of exogenous electron donors like DPC. At this stage, it is not possible to rule out the involvement of these polypeptides in the functioning of PS II-mediated electron transport.

Plastoquinone

Plastoquinone is a mobile electron carrier shuttling between PS II and the Cyt b_6/f complex (Whitmarsh, 1986). It is a relatively small molecule consisting of a quinone ring with two methyl groups and an isoprenoid chain. The size of the PQ pool has been estimated to be between 7 molecules per P_{700} (in the cyanobacterium *Anabaena variabilis*) and 40 molecules per P_{700} (in chloroplasts and *Anacystis nidulans*) (see Whitmarsh, 1986). Each quinone molecule transfers two electrons from PS II to Cyt b_6/f complex during a reduction–oxidation reaction. This is accompanied by the uptake of two H^+ ions from the stroma (during reduction at PS II) and release of the same to the lumen (during oxidation by Cyt b_6/f complex). The flux of electrons through the PQ pool seems to be the rate-limiting step in photosynthetic electron transport. In cyanobacterial thylakoids, the PQ pool is also a component of the respiratory electron transport (Hill and Bendall, 1960).

Cytochrome b6/f

This complex functions as a membrane-embedded plastoquinone-plastocyanin oxidoreductase, catalyzing the reaction

$$PQH_2 + 2\text{Plastocyanin}_{ox} \rightarrow PQ + 2\text{Plastocyanin}_{red} + 2H^+ \tag{2.5}$$

The cytochrome $f_{556.5}$ was first purified and characterized in *Spirulina platensis* by Bohme et al. (1980). The cytochrome b_6/f complex isolated from *Spirulina* sp. was reported to have four polypeptides: cytochrome f (29 kDa) with one heme c, cytochrome b_6(23 kDa) with two protohemes, iron sulfur protein (23 kDa) with one high potential iron sulfur center, and a 17 kDa polypeptide (Minami et al., 1989). Isolation of cytochrome f by Ho and Krogmann (1980) and its characterization indicated it to be a monomer of molecular weight 38 kDa. Interestingly, even though a single band of 38 kDa was obtained on the SDS-PAGE, under non-denaturing conditions two bands were observed. Also, isoelectric focusing demonstrated heterogeneity in the population of cytochrome f complexes – a major band at pI of 4.01 and a minor band at pI of 3.97. The implications of this heterogeneity are not yet clear.

Cyanobacterial cytochrome b_6/f complex links the respiratory chain to the photosynthetic electron transfer chain (Peschek and Schmetterer, 1982; Sandmann and Malkin, 1984; Binder, 1982). Its composition is similar to the bacterial Cyt bc_1 and is much simpler than the Cbc_1 complex of mitochondria. The unique aspect of the electron transport network of Cyt b_6/f complex is the possibility of feedback loops involving electron donation from ferredoxin during cyclic electron flow around PS I (Cramer et al., 1987) and oxidation of cytoplasmic electron donors, such as sulfide, during anoxygenic photosynthesis in cyanobacteria (Shahak et al., 1987).

Electron donors to PS I : Plastocyanin and Cyt c_{553}

Plastocyanin, a copper-containing peripheral membrane protein (10.5 kDa) is localized on the luminal surface of the thylakoids (Katoh, 1977; Haehnel et al., 1980) and mediates the electron transport between Cyt b_6/f complex and PS I. However, in many cyanobacteria, the role of plastocyanin is taken over by Cyt c_{553} (10–12 kDa). Some cyanobacteria also possess the ability to synthesize both plastocyanin and Cyt c_{553} and regulate their amounts depending on the availability of iron and copper (Bricker et al., 1986).

Photosystem I

PS I catalyzes the photochemical reduction of ferredoxin using reduced plastocyanin as the source of electrons.

$$\text{Ferredoxin}_{ox} + \text{Plastocyanin}_{red} + h\nu \rightarrow \text{Ferredoxin}_{red} + \text{Plastocyanin}_{ox} \qquad (2.6)$$

Studies on PS I complexes isolated from organisms belonging to diverse groups revealed that the composition, as well as biochemical, biophysical and immuno-logical properties of PS I, have been highly conserved during evolution (Bruce and Malkin, 1988a, 1988b; Lundell et al., 1985; Nechushtai et al., 1983, 1985).

The PS I core complex contains photooxidizable Chl P_{700}, about 100 molecules of Chl a as antenna, 12 to 16 molecules of carotene, 2 molecules of phylloquinone, and three 4Fe–4S centers (Bryant, 1992; Golbeck, 1992). More than a dozen polypeptides are involved in maintaining the structure and function of PS I. Two of these, with a molecular weight of about 70 kDa each on the SDS-PAGE, comprise a heterodimer which binds the entire chlorophyll present in the PS I core complex along with all the cofactors, i.e. P_{700}, A_o, A_1, 4Fe–4S clusters, etc. (Golbeck, 1992).

In PS I, charge separation starts with the photooxidation of the pigment P_{700} which is a Chl a dimer (Norris et al., 1971; Golbeck and Bryant, 1991). The electron moves sequentially through A_o, shown to be a Chl a molecule (Shuvalov et al., 1986), and A_1, a vitamin K molecule, to the first 4Fe–4S cluster, Fx. From Fx the electron is transferred to two other iron–sulfur clusters, shown to be 4Fe–4S by Scheller et al. (1989) and Petrouleas et al. (1989) coordinated on subunit C on the stromal side of PS I. Finally, ferredoxin (Fd) takes over the electron for transport to $NADP^+$ reductase.

Spirulina platensis (Arthrospira)

Attempts to purify the PS I complex from *Spirulina* led to isolation of two kinds of complexes with a molecular weight of 140 kDa and 320 kDa, under conditions of non-denaturing polyacrylamide gel electrophoresis (PAGE). The polypeptide composition of these complexes was found to be similar and each had Chl:P_{700} ratio of 100–110. It was inferred, therefore, that one of the complexes was a trimer of the basic PS I complex (Shubin et al., 1992). Surprisingly, the monomeric and trimeric forms were spectrally distinct. It was argued that the oligomerization of the cyanobacterial PS I units occurs during its isolation and may not reflect the *in vivo* condition. However, the spectral characteristics (fluorescence emission, circular dichroism and absorbance) of the thylakoid membranes of *Spirulina* sp. closely resemble that obtained from the purified trimeric complexes and differ distinctly from the monomeric PS I units (Shubin et al. 1993), suggesting that the trimeric form of PS I units already exists and is predominant in these membranes.

Recently, the three-dimensional structure of PS I, isolated from *Synechococcus* sp., has been elucidated (Krauss et al., 1993) by X-ray crystallography. As also observed for *Spirulina*, the crystals were in the form of PS I trimers. Electron densities corresponding to various cofactors of the electron transport chain of PS I were shown to be arranged on the central axis defined by the heterodimer. The electron densities were sequentially arranged, from luminal to the stromal side, for P_{700}, A_o, A_1, 4Fe–4S centers, etc. as expected for electron migration from the luminal side to the stromal side of PS I complex.

ATP synthase

The transport of electrons from PS II to PS I is accompanied by a vectorial transport of H^+ from stroma to the lumen of the thylakoids. In addition, the protons are also released into the lumen due to oxidation of H_2O by PS II. The electrochemical gradient thus generated is utilized by the ATP synthase of the thylakoids. The ATP synthase comprises two morphologically and functionally distinct parts. The hydrophilic part, F_1, emerges from the membrane and harbors the catalytic sites responsible for ATP synthesis and/or hydrolysis. Based on the native PAGE, the molecular weight of the holoenzyme (CF_1) was determined to be 320 kDa. It consists of 5 subunits α, β, δ and ε of molecular weights 53.4 kDa, 51.6 kDa, 36 kDa, 21.1 kDa and 14.7 kDa, respectively (Hicks and Yocum, 1986a). The exact stoichiometry of these subunits has not yet been determined for the *Spirulina* sp. ATPase. The other part, F_o, is embedded in the lipid bilayer and acts as a proton channel; it comprises 4 subunits: I, II, III, IV.

An interesting feature of cyanobacterial ATPases is their ability to synthesize ATP on account of the proton motive force generated by both photosynthetic and respiratory electron transport. Investigations have been carried out to determine whether these ATPases resemble mitochondria type or chloroplast type ATPases. Studies with *Spirulina* sp. ATPase suggest that it resembles the chloroplast ATPase in many ways. The magnitude of the proton motive force generated during illumination is of the same order as found in chloroplasts leading to an acidic interior (lumen) (Padan and Schuldiner, 1978). As is characteristic of chloroplast ATPase, the ATPase activity of *Spirulina* sp. membranes and isolated F_1 ATPase is mostly latent, and may be elicited by treatment with trypsin or methanol. These ATPases also exhibit magnesium-dependence of light-induced activity and calcium-dependence of trypsin-activated CF_1 activity (Hicks and Yocum, 1986b).

26

Primary sequence analysis shows broad structural and sequence similarity between chloroplast ATP synthase and those from bacteria and mitochondria. Reconstitution studies involving subunit components from diverse organisms, higher plants, and cyanobacteria and bacteria, lend credence to evolutionary relatedness (Bar-Zvi et al., 1985; Hicks et al., 1986; Richter et al., 1986; Gatenby et al., 1988).

Interaction of the Photosynthetic and Respiratory Electron Transport Chains

In several cyanobacterial sp., a light induced inhibition of respiratory oxygen consumption was observed (Brown and Webster, 1953). Such an inhibition was DCMU insensitive and was promoted by the presence of far-red light (Hoch et al., 1963). To explain such a phenomenon, a common segment was proposed to exist between the respiratory and photosynthetic electron transport chains (Jones and Myers, 1963). The electron carriers suggested to be common to both the electron transport chains are PQ (Hirano et al., 1980), plastoquinol Cyt c_{553} oxido-reductase (Peschek, 1983) and plastocyanin or Cyt c_{553} (Lockau, 1981). However, the respiratory chain has also been proposed to be located in the plasma membrane, rendering the existence of a common segment very unlikely. Gonzalez de la Vara and Gomez-Lojero (1986) investigated the possibility of such common oxido-reductants in *Spirulina maxima* since it shows a high rate of respiration as compared with other cyanobacteria.

These studies indicated an extensive overlap between the photosynthetic and respiratory electron transport chain. NADPH could be oxidized both in the dark and in the light. Furthermore, the K_m for NADPH dehydrogenase determined in the dark was similar to the K_m obtained in the light, suggesting that NADPH H^+ was oxidized in both the dark and the light by the same enzyme. Plastoquinone was found to be involved in dark NADPH oxidation and hence is considered to be a part of the respiratory chain as well. Cyt c_{553} was found to be able to reduce both Cyt oxidase

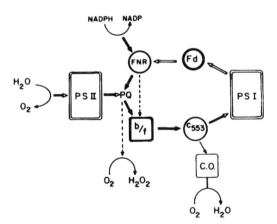

Figure 2.3 Proposed scheme for the electron transport in thylakoids of *Spirulina maxima*. Solid arrows denote reactions common to the photosynthesis and respiration; open arrows and thin arrows represent photosynthetic reactions and respiratory reactions, respectively. Broken lines indicate unknown oxido-reductants. b/f, Cyt b_6/f complex; c_{553}, Cyt c_{553}; c.o., Cyt oxidase (reproduced with permission from Gonzalez de la Vara and Gomez-Lojero, 1986).

and P_{700} (of photosystem I). FNR (ferredoxin $NADP^+$ reductase) was shown to act as the NADPH dehydrogenase in the dark. Based on these observations, a scheme of electron transport in the *Spirulina maxima* thylakoids was proposed (Figure 2.3) which demonstrated the possible interaction of respiratory and photosynthetic chains.

CO₂ Fixation

The CO_2 fixation in cyanobacteria follows the C_3-metabolism of higher plants. The Calvin cycle of CO_2 fixation seems to be operative in *Spirulina* sp. In most species of *Spirulina* a large amount of RUBISCO (ribulose bisphosphate carboxylase/oxygenase) is present, accounting for the low affinity for CO_2. In laboratory cultures, a high amount of bicarbonate is added to account for the low pCO_2 in the alkaline medium. This growth condition promotes oxygenase activity of RUBISCO and formation of glycolate, which is secreted into the medium.

Spectral Characteristics

Absorption

The intact trichomes of *Spirulina platensis* shows similar absorption characteristic (Figure 2.4A) to that of unicellular *Synechococcus* sp. The red absorption maximum due to the absorption of Chl *a* is observed at 678 nm, while the peak at 622 nm is due to absorption by PC. The relative ratio of PC absorption and the Chl *a* absorption peak vary depending on growth conditions. The soret band due to Chl *a* absorption usually occurs at 440 nm in *Spirulina* sp., while the 490 nm hump in the blue region of the spectrum is due to the absorption of carotenoids (Fork and Mohanty, 1986). The room temperature absorption spectrum of intact PBsomes isolated from *Spirulina platensis* typically exhibits a peak at 615 nm, due to PC absorption, with a conspicuous shoulder at 652 nm (APC absorption). At 77 K, the absorption spectrum exhibits three distinct peaks occurring at 590 nm, 630 nm and 650 nm (Figure 2.5). The 630 nm absorption band arises from PC absorption, while the 650–655 nm absorption is due to APC. The 590 nm peak in the 77 K absorption spectrum has been shown to occur on account of a phycobiliviolin type of chromophore attached to the α subunit of PC (Babu et al., 1991). The PC to APC ratio in *Spirulina platensis* is approximately 2.9 to 3.0, varying considerably depending on the spectral quality and intensity of light during growth. Thylakoids virtually free of PBsomes can be separated from *Spirulina* trichomes by breaking the cells with French Press. These thylakoids show a peak absorption at 678 nm due to Chl *a*, along with its soret band at 438 nm. A predominant hump at 482 nm arises from carotenoid absorption (Shubin et al., 1991).

Fluorescence

Depending on the excitation wavelength, the room temperature fluorescence emission spectrum of intact blue-green algal cells exhibits various extents of contribution of PBsome emission to the spectrum. If one exclusively excites Chl *a*,

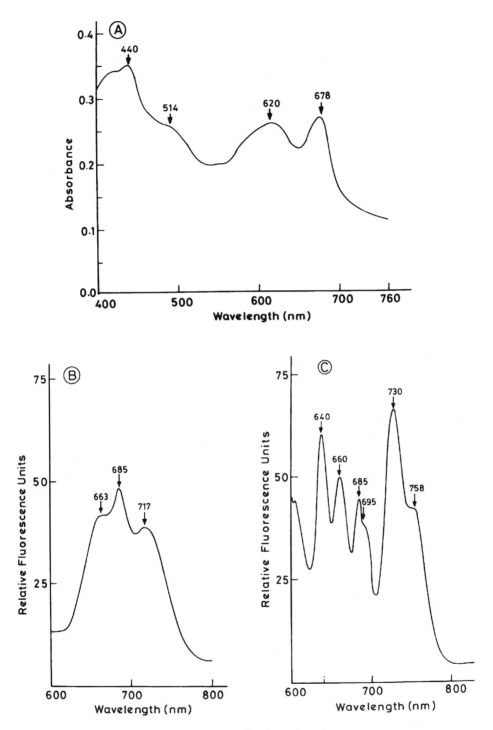

Figure 2.4 Spectral characteristics of intact cells of *Spirulina platensis*. A, room temperature absorption spectrum; B and C, fluorescence emission spectra recorded at room temperature and at 77 K, respectively. The cells were excited at 440 nm. For the 77 K spectrum, cells were suspended in 60 per cent glycerol.

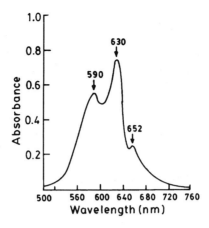

Figure 2.5 Absorption spectra of *Spirulina platensis* PBsomes recorded at 77 K.

using a monochromatic 442 nm line of an He–Cd laser, the emission spectrum by cyanobacterial cells shows no appreciable emission of PC or APC; but with broad-band blue excitation, emission from both PBsomes and Chl *a* was observed (Figure 2.4B). With the excitation of cells in the PC absorption region at 540 nm, the emission spectrum exhibits a peak at 683 to 685 nm due to Chl *a* emission and at 655 nm due to (PC + APC) emission. As in other cyanobacteria, the addition of the herbicide DCMU to *Spirulina*, which inhibits electron flow at Q_A, increases the yield of both 685 nm emission and 655 nm emission. The excitation spectrum of blue-green algae measured at 700 nm is dominated by PC 'activity' while Chl *a* 'activity' is low (Goedheer, 1968; Govindjee and Mohanty, 1972). The excitation spectrum shows a small excitation band at 435 nm (Chl *a*) and an excitation peak at 625 nm due to PC contribution (Fork and Mohanty, 1986). Detailed analyses of *Spirulina* sp. excitation characteristics have not been made. It would be of interest to study the excitation characteristics of this interesting cyanobacterium as a function of temperature. At 77 K, the emission bands become well-resolved. With excitation of blue light, the emission bands at 645, 685, 695 and 729 nm are well-resolved. In *Spirulina* sp., a far-red Chl *a* emission band at 752–758 nm is also seen. The 645 nm emission originates from PC, the 685 and 695 nm emission bands originate from PS II, and the 729 nm band emanates from PS I. Excitation of both PBsomes and core Chl *a* antenna of PS II produces a similar emission spectrum, except that the PC excitation produces a higher yield of F_{645} band.

Energy transfer characteristics of PBsomes and subunits

The energy transfer within PBsomes and from PBsomes to Chl a_2 of PS II has been well-characterized (Glazer et al., 1985; Holzwarth et al., 1983). No specific report on the energy transfer characteristics of *Spirulina* sp. PBsome is available. We assume that the energy transfer within PBsomes of *Spirulina* sp. is similar to that in other cyanobacteria. Briefly, the excitation energy absorbed by PBsomes is transferred to the PS II core antenna with about 80 to 90 per cent efficiency. With the use of time-resolved fluorescence and absorption spectroscopy, it has been shown that the chromophores within the PBsome subunits are so ordered that homotransfer is minimized and energy flows unidirectionally towards the PBsome terminal emitter, i.e. from PC→APC→APCβ→Chl a_2 (Glazer et al., 1985; Mimuro and Fujita, 1977). According to Glazer et al. (1985), the excitation energy transfer occurs from one disk to another within the rod structure of PBsomes. Inter-rod energy transfer is insignificant, thus making energy flow within the rod directional to the final emitter in the core of the PBsomes.

With the absence of any data on the architecture of PBsomes of *Spirulina* sp., it would be risky to postulate another mode of energy transfer other than the general pattern described above. In general, the PBsomes transfer their energy to the core antenna of PS II. Only under special conditions do the PBsomes transfer energy to PS I Chl a (Mullineaux, 1992). This aspect is not yet fully understood.

PS II fluorescence characteristics: Variable fluorescence

At room temperature, the bulk of the fluorescence is emitted by PS II (Goedheer, 1968). All dark-adapted oxygen-evolving photosynthetic organisms during illumination exhibit polyphasic changes in Chl a fluorescence intensity, called the Kautsky transient (Kautsky and Hirsch, 1931). The OIDPS type of Chl a fluorescence of variable yield has been observed in cyanobacteria (Mohanty and Govindjee, 1973a). Blue-green algae are also known to show fluorescence yield changes, even in the presence of the herbicide DCMU (Mohanty and Govindjee, 1973b; Papageorgiou, 1975). Detailed analysis of Chl a fluorescence of variable yield has not been made in *Spirulina* sp. However, a Chl a fluorescence yield analysis using a pulse amplitude-modulated fluorometer has been carried out for *Spirulina* trichomes immobilized on filter papers (Sivak and Vonshak, 1988). This technique provided easy analysis of photochemical and non-photochemical quenching (Schreiber et al., 1986). It was shown that the age of the culture and nutrient supply influenced photochemical quenching. The variable fluorescence yield of Chl a in cyanobacteria was low because of the high F_0 level. Like Chl a fluorescence, the fluorescence of PC and APC undergoes slow, time-dependent changes. This is assumed to be due to reverse energy transfer from Chl a_2 to PBsome (Fork and Mohanty, 1986). The abundance of Chl a in PS I, APC fluorescence and the reverse energy transfer all possibly contribute to a high level of F_0 in cyanobacteria.

It is known that the PC to APC ratio controls the energy transfer from PBsomes to Chl a_2 (Fork and Mohanty, 1986). In *Spirulina platensis* grown in continuous light under laboratory conditions, this ratio is about 2.8 to 3.0. The PC to APC ratio might affect back energy transfer. PC to APC ratio is variable in *Spirulina* sp. (Babu et al., 1991), depending on growth conditions.

The PS I fluorescence characteristics

The bulk of Chl a in cyanobacteria resides in PS I (Mimuro and Fujita, 1977), but at room temperature the PS I Chl a is only weakly fluorescent. Fluorescence yield of PS I increases when the temperature is lowered to 77 K or below. At 77 K, a fluorescence band at 730 nm (F_{730} originating from PS I) dominates the spectrum. The fluorescence yield of the F_{730} band at 77 K is variable with time of illumination, the variable yield of PS I Chl a fluorescence being linked to a spillover of energy from PS II to PS I (Butler, 1978). In *Spirulina* sp., besides this F_{730} emission band,

31

an additional long-wavelength Chl a emission band at 756–760 nm (F_{758}) has been observed at 77 K (Shubin et al., 1991). This long-wavelength emission band is very pronounced in *Spirulina platensis* (Figure 2.6) and originates from a long-wavelength form of Chl a, designated as a_{735}, which constitutes approximately 5 per cent of the absorption of the Chl a red absorption band. F_{758} gets photobleached, and this photobleaching is linked to the redox state of P_{700}, the physiological donor of PS I (Shubin et al., 1991).

The variable yield of F_{758} at 77 K is dependent on the redox state of P_{700}, and the quantum yield of F_{758} is proportional to the P_{700} concentration (Shubin et al., 1991). The photooxidation of P_{700} at 77 K possibly quenches the $F_{758-760}$ due to energy transfer from Chl a_{735} to $P_{700}{}^+$. Thus, this long-wavelength Chl a fluorescence seems to provide a probe for monitoring stress damage to PS I in *Spirulina* (Babu et al., 1992). Two types of Chl a–protein complexes have been isolated from *Spirulina platensis* using Triton X-100 (Shubin et al., 1992). The high molecular weight (320 kDa) complex contains the long-wavelength form of Chl a_{735}. It contains trimeric chlorophyll–protein complex of PS I and seems to occur *in vivo* in *Spirulina* sp. (Shubin et al., 1993).

Energy Transfer and Spillover

The stoichiometry of PS II to PS I is variable. It is also known that the ratio of PBsomes to the PS II reaction center is variable in the range 1 to 4 depending on the alga and growth conditions (Fork and Mohanty, 1986). The bulk of excitation energy gets transferred from PBsomes to PS II, and more than one PBsome seems to feed energy to one PS II. The Chl a content of the PS II core antenna is smaller than that

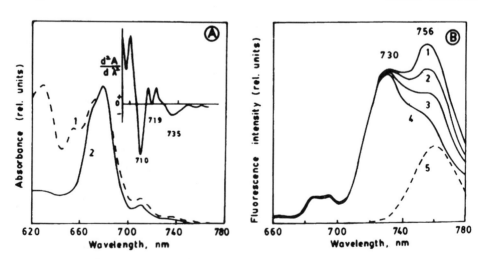

Figure 2.6 77 K absorbance and fluorescence spectra of *Spirulina platensis* in 60 per cent glycerol. (A) absorption spectra of whole cells (1) and membranes (2). Inset: the second derivative of absorption spectrum (2) in the red region. (B) fluorescence spectra of whole cells with excitation at 435 nm. Cells were frozen in the dark and spectra were measured a| 5 (1), 20 (2) or 80 (3) mW m^{-2} intensity of exciting light, or cells were frozen in monochromatic exciting light at 20 mW m^{-2} (4); the spectra were normalized at 720 nm. (5) curve 1 minus curve 4 (reproduced with permission from Shubin et al., 1991).

of PS I; the excitation of Chl *a*, therefore, preferentially excites PS I. Diner (1979) showed that in the red alga *Cyanidium* not all the PBsomes are attached to the PS II, and it is probable that a similar situation occurs in cyanobacteria. It is not known if these PS II-delinked PBsomes transfer energy to PS I. Such transfer has been shown to occur in *Synechococcus* 6301 (Mullineaux, 1992). The extent of spillover from Chl *a* of PS II to Chl *a* of PS I has not been estimated in *Spirulina* sp.

A detailed investigation of the spectral features and energy transfer characteristic of *Spirulina* has not been carried out. However, this trichotomous alkalophilic cyanobacterium appears to be good test material for studies relating to specific changes in energy transfer associated with environmental stress.

Dynamic Nature of the Photosynthetic Apparatus

Photosynthetic organisms are frequently exposed to many stress conditions that lead to an imbalance in the electron transport rate through the two photosystems. Such stress factors could be environmental and genetic, including alterations in the quality and quantity of light, nutritional stress, alterations in the size of antenna due to mutations, etc. Natural-based self-shading as well as depth of canopy favour a higher PSI absorption, leading to an imbalance in energy absorption by the two photosystems. The structure and function of the thylakoid membrane is altered under stress conditions (Melis, 1991; Melis et al., 1985). The alterations are both short- and long-term, depending on the nature and duration of the stress.

Short-term Changes

The phenomenon by which photosynthetic organisms achieve a redistribution of excitation energy between the two photosystems under light conditions that overexcite either PS I or PS II has been termed 'state transition'. Overexcitation of PS I establishes State 1 where the available light energy is distributed in favor of PS II. During State 2, established by overexcitation of PS II, a higher proportion of energy becomes available to PS I. Such a redistribution of energy occurs on the order of a few minutes and is purely reorganizational in nature (Fork and Satoh, 1986). The phenomenon of state transitions has not been investigated in *Spirulina* sp. However, studies on other cyanobacteria like *Synechococcus* sp. and *Synechocystis* sp. have provided considerable, although not complete, understanding of this phenomenon (Biggins and Bruce, 1989; Allen, 1992).

Compared with the higher plants, the state transitions in cyanobacteria have a faster kinetics. Also the effect of state transitions on the redistribution of excitation energy is much more marked (Fork and Satoh, 1986). Both the cyclic electron flow around PS I (Satoh and Fork, 1983) and the redox state of the PQ pool (Mullineaux and Allen, 1990) have been proposed to be the primary signals for redistribution of excitation energy during state transitions. Phosphorylation of 18.5 kDa (weakly membrane-bound) and 15 kDa (membrane-bound) proteins has been suggested to play an important role in state transitions, correlating with the State 2 transition in *Synechococcus* 6301. The two proteins are components of PBsomes and PS II core (Allen et al., 1985). However, some studies have indicated that the phosphorylation of PBsomes components is not necessary for the state transitions in cyanobacteria

(Biggins and Bruce, 1987; Bruce et al., 1989) since even a phycobilisome-less mutant of *Synechococcus* 7002 did retain the ability to undergo such transitions. Three mechanisms have been proposed that affect the redistribution of energy:

1   Energetic decoupling of PBsomes and PS II during State 2 reduces the amount of energy available for PS II without affecting the energy available for PS I photochemistry (Mullineaux and Holzwarth, 1990; Mullineaux et al. 1990).

2   Phycobilisome transfers energy preferentially to PS II during State 1 and to PS I during State 2 (Mullineaux, 1992).

3   The extent of energy transfer from PS II to PS I (spillover) is higher in State 2 than State 1 and is governed by the proximity of PS II and PS I (Bruce et al., 1985; Bruce and Salehian, 1992).

Thus, none of the currently proposed models for the state transitions in cyanobacteria can fully explain the details of the three aspects of such transitions. The signal for the transition, mediation of the signal, and the final reorganization of the photosynthetic apparatus that effects a redistribution of excitation energy are all presently controversial. It is worth noting that state transitions are also influenced by the intensity of polychromatic white light. At low PAR (photosynthetically active radiation), cells adapt to a low fluorescent State 2, while at high PAR, State 1 is established (Rouag and Dominy, 1994).

Long-term Changes

A variety of environmental, developmental and genetic factors may cause prolonged stress to the photosynthetic apparatus. In such a situation, the composition of the thylakoids becomes altered. This process involves alterations in the rate of biosynthesis, assembly and degradation of various supramolecular complexes of the thylakoids. Antenna sizes of the two photosystems and relative stoichiometry of the complexes could become altered during this process of acclimation (Melis, 1991). Further, these changes can be reversed if the stress is alleviated.

The most important stress condition experienced by photosynthetic organisms is the variation in the intensity of light and its spectral quality. Excessively high intensity of light has been shown to cause photoinhibition in *Spirulina platensis* (Vonshak et al., 1988; Jensen and Knutsen, 1993), as found in other cyanobacteria and higher plants. The relative content of phycocyanin, chlorophyll and carotenoids has been observed to change under differing conditions of prevalent light (Shyam and Sane, 1989), possibly to prevent/reduce the photoinhibitory damage.

In addition, the intensity of light during growth seems to regulate three different aspects of the photosynthetic apparatus:

●   The antenna sizes of both PS I and PS II are larger in plants growing in low light. PS II antenna size is affected to a greater extent. This has been independently shown by Lichtenthaler et al. (1982), Hodges and Barber (1983) and Leong and Anderson (1983, 1984a, 1984b) for higher plants. In the case of cyanobacteria, the size of PBsomes (antenna for PS II) increases under low light irradiance (Glazer, 1984).

●   The PS II:PS I stoichiometry decreases under low light intensity in cyanobacteria (Kawamura et al. 1979; Murakami and Fujita, 1991) and in chloroplasts from several species (Melis and Harvey, 1981; Wild et al., 1986; Leong and Anderson 1986).

- The content of the Cyt b_6/f complex is higher relative to the reaction centers when the photosynthetic organisms are grown at a higher intensity of light (Melis, 1991). Hence, its content may be considered to reflect the electron transport rate through the two photosystems (Wilhelm and Wild, 1984).

The action spectrum of the two reaction centers is quite different. The PBsomes have absorption properties very distinct from those of chlorophyll, and the energy harvested by them is preferentially transferred to PS II (Ley and Butler, 1980). The light-harvesting antenna of PS I in all photosynthetic organisms has a predominance of Chl *a*. Such major differences in the action spectrum of the two photosystems imply that the quality of irradiating light has the potential to exert a very strong effect on the relative activity of the two photosystems.

Cyanobacteria grown in light that overexcites PS II (yellow light) have a lower PS II to PS I ratio as compared with those grown in light which overexcites PS I (far-red light). This was clearly shown by Myers et al. (1980), Manodori and Melis (1986), and Fujita and Murakami (1987). The antenna sizes for PS II and PS I seem to remain unaltered under these conditions.

The composition of the PBsomes in terms of relative amounts of PC and phycoerythrin may be altered depending on the ambient light conditions. The phenomenon is termed complementary chromatic adaptation and has been recently reviewed (Tandeau de Marsac, 1991; Grossman et al., 1993). Due to the absence of phycoerythrin, *Spirulina* does not undergo typical chromatic acclimation, but the content of PXB, phycobiliviolin type of chromophore has been observed to increase in *Spirulina platensis* grown in green light (Babu et al., 1991).

Although the response of the photosynthetic apparatus of *Spirulina* towards different stress conditions has not been investigated in detail, *Spirulina* could be expected to have the same dynamic ability to overcome prevailing stress as other photosynthetic organisms by altering effective antenna size and the ratio of PS I to PS II.

Conclusions

There is a growing need for investigation of the basic aspects of photosynthesis in the blue-green alga *Spirulina*, to ensure optimal biomass production, because of its growing importance in rural biotechnology and commercial production. Unfortunately, studies on photochemical reactions and the carbon metabolism of this important cyanobacterium are scant. *Spirulina* offers many unique features for studying photosynthetic apparatus, such as the dynamics of PBsomes and energy transfer within the PBsomes for optimal light harvest and utilization, and topology and heterogeneity of photosystems, particularly PS I, and energy metabolism of photosynthetic and respiratory membranes. This mesophilic cyanobacterium grows under extreme alkaline conditions. However, no work seems to have been done on its pH-stat mechanism.

Similarly, very little work has been carried out on carbon and nitrogen metabolism and their regulation. *Spirulina* is a good material for preparation of active PS II, and it has unique PS I. Although *Spirulina* sp. has been used in metal stress and photoinhibitory studies, little research has been carried out on the molecular mechanism. It appears to us that *Spirulina* would afford a good opportunity to probe photoinhibition in PS I. We hope that its unique cell-biology will encourage further investigation, regardless of its enormous commercial importance.

Acknowledgements

The authors wish to acknowledge Prof. Govindjee of the University of Illinois, USA for his help in the literature survey. The work done in the authors' laboratory was supported by grant 09/0374/EMRII from the Council of Scientific and Industrial Research (CSIR), New Delhi. MS thanks CSIR for a fellowship.

References

AKABORI, K., TSUKAMOTO, H., TSUKIHARA, J., NAGATSUKA, T., MOTOKAWA, O. and TOYOSHIMA, Y. (1988) Disintegration and reconstitution of photosystem II reaction centre core complex. I. Preparation and characterization of 3 different types of subcomplex, *Biochim. Biophys. Acta*, **392**, 345.

ALLEN, J. F. (1992) Protein phosphorylation in regulation of photosynthesis, *Biochim. Biophys. Acta*, **1098**, 275.

ALLEN, J. F., SANDERS, C. E. and HOLMES, N. G. (1985) Correlation of membrane protein phosphorylation with excitation energy distribution in cyanobacterium *Synechocystis*, 6301. *FEBS Lett.*, **193**, 271.

BABU, T. S., KUMAR, A. and VARMA, A. K. (1991) Effect of light quality on phycobilisome components of the cyanobacterium, *Spirulina platensis*, *Plant Physiol.*, **95**, 492.

BABU, T. S., SABAT, S. C. and MOHANTY, P. (1992) Heat induced alterations in the photosynthetic electron transport and emission properties of cyanobacterium *Spirulina platensis*, *J. Photochem. Photobiol.*, **12**, 161.

BAKELS, R. H. A., VAN WALRAVEN, H. S., KRAB, K., SCHOLTS, M. J. C., and KRAAYENHOF, R. (1993) On the activation mechanism of the H^+-ATP synthase and unusual thermodynamic properties in the alkalophilic cyanobacterium *Spirulina platensis*, *Eur. J. Biochem.*, **213**, 957.

BARBER, J., CHAPMAN, D. J. and TELFER, A. (1987) Characterization of a photosystem II reaction centre isolated from the chloroplasts of *Pisum sativum*, *FEBS Lett.*, **220**, 67.

BAR-ZVI, D., YOSHIDA, M. and SHAVIT, N. (1985) Reconstitution of photophosphorylation with coupling factor 1 ATPases from the thermophilic bacterium PS3 and lettuce chloroplasts, *Biochim. Biophys. Acta*, **806**, 341.

BENNETT, A. and BOGORAD, L. (1973) Complementary chromatic adaptation in a filamentous blue green alga, *J. Cell Biol.*, **58**, 419.

BIGGINS, J. and BRUCE, D. (1987) The relationships between protein kinase activity and chlorophyll *a* fluorescence changes in thylakoids from the cyanobacterium *Synechococcus* PCC 6301. In BIGGINS, J. (Ed.) *Progress in Photosynthesis Research*, Vol. 2, p. 773, Dordrecht: Martinus Nijhoff.

BIGGINS, J. and BRUCE, D. (1989) Regulation of excitation energy transfer in organisms containing phycobilisomes, *Photosynth. Res.*, **20**, 1.

BINDER, A. (1982) Respiration and photosynthesis in energy transducing membranes of cyanobacteria, *J. Bioeng. and Biomembranes*, **14**, 271.

BJORKMAN, O. and DEMMING, B. (1987) Photon yield of oxygen evolution and chlorophyll fluorescence at 77 K among vascular plants of diverse origins, *Planta*, **170**, 489.

BOHME, H., PEIZER, B. and BOGER, P. (1980) Purification and characterization of cytochrome $f_{556.5}$ from blue-green alga *Spirulina*, *Biochim. Biophys. Acta*, **592**, 528.

BRICKER, T. M., GUIKEMA, J. A., PAKRASI, H. B. and SHERMAN, L. A. (1986) Proteins of cyanobacterial thylakoids. In Staehelin, L.A. and Arntzen, C.J. (Eds) *Encyclopaedia of Plant Physiology – New Series*, Vol. 19, p. 640, Heidelberg: Springer-Verlag.

BROWN, A. H. and WEBSTER, G. C. (1953) The influence of light on the rate of respiration of the blue-green alga *Anabaena*, *American J. Bot.*, **40**, 753.

BRUCE, B. D. and MALKIN, R. (1988a) Structural aspects of photosystem I from *Dunaliella salina*, *Plant Physiol.*, **88**, 1201.

BRUCE, B. D. and MALKIN, R. (1988b) Subunit stoichiometry of the chloroplast photosystem I complex, *J. Biol. Chem.*, **263**, 7302.

BRUCE, D. and SALEHIAN, O. (1992) Laser induced optoacoustic calorimetry of cyanobacteria. The efficiency of primary photosynthetic processes in state I and state II, *Biochim. Biophys. Acta*, **1100**, 242.

BRUCE, D., BIGGINS, J., STEINER, T. and THEWALT, M. (1985) Mechanism of the light state transition in photosynthesis. IV. Picosecond fluorescence spectroscopy of *Anacystis nidulans* and *Porphyridium cruentum* in state I and state II at 77 K, *Biochim. Biophys. Acta*, **806**, 237.

BRUCE, D., BRIMBLE, S. and BRYANT, D. A. (1989) State transitions in a phycobilisome less mutant of cyanobacterium *Synechococcus* PCC 7002, *Biochim. Biophys. Acta*, **974**, 66.

BRYANT, D. A. (1991) Cyanobacterial phycobilisomes: Progress toward complete structural and functional analysis via molecular genetics. In BOGORAD, L. and VASIL, I. K. (Eds) *The Photosynthetic Apparatus – Molecular Biology and Operation*, p. 257, San Diego: Academic Press.

BRYANT, D. A. (1992) Photosystem I: Polypeptide subunits, genes and mutants. In BARBER, J. (Ed.) *Topics in Photosynthesis – The Photosystems: Structure, Function and Molecular Biology*, Vol. 11, p. 501, Amsterdam: Elsevier Science.

BUTLER, W. L. (1978) Energy distribution in the photochemical apparatus of photosynthesis, *Ann. Rev. Plant Physiol.*, **29**, 345.

CRAMER, W. A., BLACK, M. T., WIDGER, W. R. and GIRVIN, M. E. (1987) Structure and function of the photosynthetic b-c_1 and b_6-f complexes. In BARBER, J. (Ed.) *The Light Reactions*, p. 446, Amsterdam: Elsevier Science.

DEBUS, R. J., BARRY, B. A., BABCOCK, G. T. and MCINTOSH, L. (1988a) Site directed mutagenesis identifies a tyrosine radical involved in the photosynthetic oxygen-evolving system, *Proc. Natl. Acad. Sci. (USA)*, **85**, 427.

DEBUS, R. J., BARRY, B. A., SITHOLE, I., BABCOCK, G. T. and MCINTOSH, L. (1988b) Directed mutagenesis indicates that the donor to P-680$^+$ in photosystem II is tyrosine-161 of the D1 polypeptide, *Biochem.*, **27**, 9071.

DEISENHOEFER, J. and MICHEL, H. (1989) The photosynthetic reaction centre from the purple bacterium *Rhodopseudomonas viridis*, *Science*, **245**, 1463.

DINER, B. A. (1979) Energy transfer from the phycobilisomes to photosystem II reaction centers in wild type *Cyanidium caldarum*, *Plant Physiol.*, **63**, 30.

DUYSENS, L. N. M. (1952) Transfer of Excitation Energy in Photosynthesis, Ph.D. Thesis, Utrecht.

EMERSON, R. and ARNOLD, W. (1932) The photochemical reaction in photosynthesis, *J. Gen. Physiol.*, **16**, 191.

FORK, D. C. and MOHANTY, P. (1986) Fluorescence and other characteristics of blue-green algae (cyanobacteria), red algae, and cryptomonads. In GOVINDJEE, AMESZ, J. and FORK, D. C. (Eds) *Light Emission by Plants and Bacteria*, p. 451, London: Academic Press.

FORK, D. C. and SATOH, K. (1986) The control by state transitions of the distribution of excitation energy in photosynthesis, *Ann. Rev. Plant Physiol.*, **37**, 335.

FUJITA, Y. and MURAKAMI, A. (1987) Regulation of electron transport composition in cyanobacterial photosynthetic system: stoichiometry among Photosystem I and II complexes and their light harvesting antennae and Cyt b_6/f complex, *Plant Cell Physiol.*, **28**, 1547.

GANTT, E. (1988) Phycobilisomes: Assessment of the core structure and thylakoid interactions. In STEVENS, S. E. and BRYANT, D. A. (Eds) *Light Energy Transduction in Photosynthesis – Higher Plant and Bacterial Models*, p. 91, Maryland, USA: American Society of Plant Physiologists.

GARNIER, F., DUBACQ, J.-P. and THOMAS, J.-C. (1994) Evidence for a transient association of new proteins with the *Spirulina maxima* phycobilisome in relation to light intensity, *Plant Physiol.*, **106**, 747.

GATENBY, A. A., ROTHSTEIN, S. J. and BRADLEY, D. (1988) Using bacteria to analyse sequences involved in chloroplast gene expression, *Photosynth. Res.*, **19**, 7.

GLAZER, A. N. (1982) Phycobilisome: Structure and dynamics, *Annu. Rev. Microbiol.*, **36**, 173.

GLAZER, A. N. (1984) Phycobilisome, a macromolecular complex optimized for light energy transfer, *Biochim. Biophys. Acta*, **768**, 29.

GLAZER, A. N. (1985) Light harvesting by phycobilisomes, *Annu. Rev. Biophys. Biophys. Chem.*, **14**, 47.

GLAZER, A. N. (1989) Light guides directional energy transfer in photosynthetic antenna, *J. Biol. Chem.*, **264**, 1.

GLAZER, A. N., YEHI, S. W., WEBB, S. P. and CLARK, J. H. (1985) Disc to disc transfer as the rate limiting step for energy flow in phycobilisomes, *Science*, **227**, 419.

GOEDHEER, J. C. (1968) On the low temperature spectrum of blue-green and red algae, *Biochim. Biophys. Acta*, **153**, 903.

GOLBECK, J. H. (1992) Structure and function of photosystem I, *Annu. Rev. Plant Physiol. Plant Mol. Biol.*, **43**, 293.

GOLBECK, J. H. and BRYANT, D. A. (1991) Photosystem I. In LEE, C. P. (Ed.) *Current Topics in Bioenergetics*, p. 83, San-Diego: Academic Press.

GONZALEZ DE LA VARA, L. and GOMEZ-LOJERO, C. (1986) Participation of plastoquinone, cytochrome c553 and ferredoxin-NADP$^+$ oxido-reductase in both photosynthesis and respiration in *Spirulina maxima*, *Photosynth. Res.*, **8**, 65.

GOUNARIS, K., CHAPMAN, D. J. and BARBER, J. (1989) Isolation and characterization of a D1/D2/cyt b_{559} complex from *Synechocystis* PCC 6803, *Biochim. Biophys. Acta*, **973**, 296.

GOVINDJEE and EATON-RYE, J. J. (1986) Electron transfer through photosystem II acceptors: Interaction with anions, *Photosynth. Res.*, **10**, 365.

GOVINDJEE and MOHANTY, P. K. (1972) Photochemical aspects of photosynthesis in blue-green algae. In DESIKACHARY, T. V. (Ed.) *Proceedings of Symposium on Taxonomy, Biology of Blue-green Algae*, p. 171, Madras: University of Madras Press.

GROSSMAN, A. R., SCHAEFER, M. R., CHIANG, G. G. and COLLIER, J. L. (1993) Environmental effects on the light harvesting complex of cyanobacteria, *J. Bacteriol.*, **175**, 575.

HAEHNEL, W., PROPPER, A. and KRAUSE, H. (1980), Evidence for complex plastocyanin as the common electron donor of P_{700}, *Biochim. Biophys. Acta*, **593**, 384.

HANSSON, O. and WYDRZYNSKI, T. (1990) Current perceptions of photosystem II, *Photosynth. Res.*, **23**, 131.

HATTORI, A. and FUJITA, Y. (1959) Formation of phycobilin pigments in a blue-green alga *Tolupothrix tennuis* as induced by illumination with coloured lights, *J. Biochem.*, **46**, 521.

HICKS, D. B. and YOCUM, C. F. (1986a) Properties of cyanobacterial coupling factor ATPase from *Spirulina platensis*. I. Electrophoretic characterization and reconstitution of photophosphorylation, *Arch. Biochem. Biophys.*, **245**, 220.

HICKS, D. B. and YOCUM, C. F. (1986b) Properties of cyanobacterial coupling factor ATPase from *Spirulina platensis*. II. Activity of purified and membrane bound enzyme, *Arch. Biochem. Biophys.*, **245**, 230.

HICKS, D. B., NELSON, N. and YOCUM, C. F. (1986) Cyanobacterial and chloroplast F_1-ATPases: Cross reconstitution of phosphorylation and subunit immunological relationships, *Biochim. Biophys. Acta*, **851**, 217.

HILL, R. and BENDALL, F. (1960) Function of the two cytochrome components in chloroplasts. A working hypothesis, *Nature*, **186**, 136.

HIRANO, M., SATOH, K. and KATOH, S. (1980) Plastoquinone as a common link between photosynthesis and respiration in blue-green algae, *Photosynth. Res.*, **1**, 149.

HO, K. K. and KROGMANN, D. W. (1980) Cytochrome f from spinach and cyanobacteria, *J. Biol. Chem.*, **255**, 3855.

HOCH, G., OWENS, O. H. and KOK, B. (1963) Photosynthesis and Respiration, *Arch. Biochem. Biophys.*, **101**, 171.

HODGES, M. and BARBER, J. (1983) Photosynthetic adaptation of pea plants grown at different light intensities: State I–state II transitions and associated chlorophyll fluorescence changes, *Planta*, **157**, 166.

HOLZWARTH, A. R., WENDLER, J. and WEHREMEYER, W. (1983) Studies on chromophore coupling in isolated phycobiliproteins. I. Picosecond fluorescence kinetics and energy transfer in phycocyanin from *Chroomonas* sp., *Biochim. Biophys. Acta*, **724**, 388.

JENSEN, S. and KNUTSEN, G. (1993) Influence of light and temperature on photoinactivation of photosynthesis in *Spirulina platensis*, *J. Appl. Phycology*, **5**, 495.

JONES, L. W. and MYERS, J. A. (1963) Common link between photosynthesis and respiration in a blue-green alga, *Nature*, **199**, 670.

KATOH, S. (1977) Plastocyanin. In TREBST, A. and AVRON, M. (Eds) *Encyclopaedia of Plant Physiology – New Series*, Vol. 5, p. 247, Berlin: Springer-Verlag.

KAUTSKY, H. and HIRSCH, A. (1931) Neul Versuche zur Kohlensaureassimilation, *Naturwissenschaften*, **48**, 964.

KAWAMURA, M., MIMURO, M. and FUJITA, Y. (1979) Quantitative relationship between the two reaction centres in the photosynthetic system of blue-green algae, *Plant Cell Physiol.*, **20**, 697.

KRAUSS, N., HINRICHS, W., WITT, I., FROMME, P., PRITZKOW, W., DAUTER, Z., BETZEL, C., WILSON, K. S., WITT, H. T. and SAENGER, W. (1993) Three dimensional structure of system I of photosynthesis at 6 Å resolution, *Nature*, **361**, 326.

LEONG, T. Y. and ANDERSON, J. M. (1983) Changes in composition and function of thylakoid membranes as a result of photosynthetic adaptation of chloroplasts from pea plants grown under different light conditions, *Biochim. Biophys. Acta*, **723**, 391.

LEONG, T. Y. and ANDERSON, J. M. (1984a) Adaptation of thylakoid membranes of pea chloroplasts to light intensities. I. Study on the distribution of chlorophyll protein complexes, *Photosynth. Res.*, **5**, 105.

LEONG, T. Y. and ANDERSON, J. M. (1984b) Adaptation of the thylakoid membranes of pea chloroplasts to light intensities. II. Regulation of electron transport capacities, electron carriers, coupling factor (CF_1) activity and rates of photosynthesis, *Photosynth. Res.*, **5**, 117.

LEONG, T. Y. and ANDERSON, J. M. (1986) Light quality and irradiance adaptation of the composition and function of pea thylakoid membranes, *Biochim. Biophys. Acta*, **850**, 57.

LERMA, C. and GOMEZ-LOJERO, C. (1982) Photosynthetic phosphorylation by a membrane preparation of the cyanobacterium *Spirulina platensis*, *Biochim. Biophys. Acta*, **680**, 181.

LEY, A. C. and BUTLER, W. L. (1980) Energy distribution in the photochemical apparatus of *Porphyridium cruentum* in state I and state II, *Biochim. Biophys. Acta*, **592**, 349.

LICHTENTHALER, H. K., KUHN, G., PRENZEL, U., BUSCHMANN, C. and MEIER, D. (1982) Adaptation of chloroplast ultrastructure and of chlorophyll protein levels to high light and low light growth conditions, *Z. Naturforsch.*, **37C**, 464.

LOCKAU, W. (1981) Evidence for a dual role of cytochrome c_{553} and plastocyanin in photosynthesis and respiration of the cyanobacterium *Anabaena variabilis*, *Arch. Microbiol.*, **128**, 336.

LUNDELL, D. J., GLAZER, A. N., MELIS, A. and MALKIN, R. (1985) Characterization of a cyanobacterial photosystem I complex, *J. Biol. Chem.*, **260**, 646.

MANODORI, A. and MELIS, A. (1986) Cyanobacterial acclimation to photosystem I or photosystem II light, *Plant Physiol.*, **82**, 185.

MELIS, A. (1991) Dynamics of photosynthetic membrane composition and function, *Biochim. Biophys. Acta*, **1058**, 87.

MELIS, A. and HARVEY, G. W. (1981) Regulation of photosystem stoichiometry, chlorophyll *a* and chlorophyll *b* content and relation to chloroplast ultrastructure, *Biochim. Biophys. Acta*, **637**, 138.

MELIS, A., MANODORI, A., GLICK, R. E., GHIRARDI, M. L., McCAULEY, S. W. and NEALE, P. J. (1985) The mechanism of photosynthetic membrane adaptation to environmental stress conditions: A hypothesis on the role of electron transport capacity and of ATP/NADPH pool in the regulation of thylakoid membrane organization and function, *Physiologie Vegetale*, **23**, 757.

MIMURO, M. and FUJITA, Y. (1977) Estimation of Chl *a* distribution in the photosynthetic pigment system I and II of blue-green alga *Anabaena variabilis*, *Biochim. Biophys. Acta*, **459**, 376.

MINAMI, Y., WADA, K. and MATSUBARA, H. (1989) The isolation and characterization of a cytochrome b6f complex from cyanobacterium *Spirulina* species, *Plant Cell Physiol.*, **30**, 91.

MOHANTY, P. and GOVINDJEE (1973a) Light induced changes in the fluorescence yield of chlorophyll *a* in *Anacystis nidulans*. I. Relationship of slow fluorescence changes with structural changes, *Biochim. Biophys. Acta*, **305**, 95.

MOHANTY, P. and GOVINDJEE (1973b) Light induced changes in the fluorescence yield of chlorophyll *a* fluorescence in *Anacystis nidulans*. II. The fast changes and the effect of photosynthetic inhibitors on both fast and slow fluorescence induction, *Plant Cell Physiol.*, **14**, 611.

MOHANTY, P., HOSHINA, S. and FORK, D. C. (1985) Energy transfer from phycobilin to chlorophyll *a* in heat treated cells of *Anacystis nidulans*. Characterization of 683 nm emission band, *Photochem. Photobiol.*, **41**, 589.

MULLINEAUX, C. W. (1992) Excitation energy transfer from phycobilisomes to photosystem I in a cyanobacterium, *Biochim. Biophys. Acta*, **1100**, 285.

MULLINEAUX, C. W. and ALLEN, J. F. (1990) State I to state II transitions in the cyanobacterium *Synechococcus* PCC 6301 are controlled by the redox state of electron carriers between Photosystem I and Photosystem II, *Photosynth. Res.*, **23**, 297.

MULLINEAUX, C. W. and HOLZWARTH, A. R. (1990) A proportion of photosystem II core complexes are decoupled from phycobilisomes in light state 2 in the cyanobacterium *Synechococcus* PCC 6301, *FEBS Lett.*, **260**, 245.

MULLINEAUX, C. W., BITTERSMANN, E., ALLEN, J. F. and HOLZWARTH, A. R. (1990) Picosecond time resolved fluorescence emission spectra indicate decreased energy transfer from phycobilisomes to photosystem II in light-state 2 in cyanobacterium *Synechococcus* 6301, *Biochim. Biophys. Acta*, **1015**, 231.

MURAKAMI, A. and FUJITA, Y. (1991) Regulation of stoichiometry in the photosynthetic system of the cyanophyte *Synechocystis* PCC 6714 in response to light intensity, *Plant Cell Physiol.*, **32**, 223.

MURTHY, S. D. S. (1991) Studies on Bioenergetic Processes of Cyanobacteria: Analysis of the Selected Heavy Metal Ions on Energy Linked Processes, Ph.D. Thesis, Jawaharlal Nehru University, New Delhi.

MURTHY, S. D. S. and MOHANTY, P. (1995) Action of selected heavy metal ions on the photosystem II activity of cyanobacterium *Spirulina platensis*, *Biologia Plantarum*, **37**, 79.

MYERS, J., GRAHAM, J.-R. and WANG, R. T. (1980) Light harvesting in *Anacystis nidulans* studied in pigment mutants, *Plant Physiol.*, **66**, 1144.

NANBA, O. and SATOH, K. (1987) Isolation of a photosystem II reaction centre consisting of D1 and D2 polypeptides and Cytochrome b_{559}, *Proc. Natl. Acad. Sci.* (USA), **84**, 109.

NECHUSHTAI, R., MUSTER, P., BINDER, A., LIVEANU, V. and NELSON, N. (1983) Photosystem I reaction centre from the thermophilic cyanobacterium *Mastigocladus laminosus*, *Proc. Natl. Acad. Sci.* (USA), **80**, 1179.

NECHUSHTAI, R., NELSON, N., GONEN, O. and LEVANON, H. (1985) Photosystem I reaction centre from *Mastigocladus laminosus*. Correlation between reduction state of the iron–sulphur centres and the triplet formation mechanisms, *Biochim. Biophys. Acta*, **807**, 355.

NORRIS, J. R., UPHAUS, R. A., CRESPI, H. G. and KATZ, J. J. (1971) Electron spin resonance of chlorophyll and the origin of signal 1 in photosynthesis, *Proc. Natl. Acad. Sci.* (USA), **68**, 625.

PADAN, E. and SCHULDINER, S. (1978) Energy transduction in the photosynthetic membranes of the cyanobacterium (blue-green alga) *Plectonema boryanum*, *J. Biol. Chem.*, **253**, 3281.

PAPAGEORGIOU, G. C. (1975) Chlorophyll fluorescence: An intrinsic probe of photosynthesis. In GOVINDJEE (Ed.) *Bioenergetics of Photosynthesis*, p. 320, New York: Academic Press.

PESCHEK, G. A. (1983) The cytochrome b/f complex – A common link between photosynthesis and respiration in cyanobacterium *Anacystis nidulans*, *Biochem. J.*, **210**, 269.

PESCHEK, G. A. and SCHMETTERER, G. (1982) Evidence for plastoquinol-cytochrome f/b_{563} reductase as a common electron donor to P_{700} and cytochrome oxidase in cyanobacteria, *Biochem. Biophys. Res. Communications*, **108**, 1188.

PETROULEAS, V., BRAND, J. J., PARETT, K. G. and GOLBECK, J. M. (1989) A Mossbaur analysis of the low potential iron–sulphur center in photosystem I: Spectroscopic evidence that Fx is a [4Fe–4S] cluster, *Biochem.*, **28**, 8980.

RICHTER, M. L., GROMET-ELHANAN, Z. and MCCARTY, R. E. (1986) Reconstitution of the H^+-ATPase complex of *Rhodopseudomonas rubrum* by the β subunit of chloroplast coupling factor, *J. Biol. Chem.*, **261**, 12109.

ROUAG, D. and DOMINY, P. (1994) State adaptations in the cyanobacterium *Synechococcus* 6301 (PCC): Dependence on light intensity or spectral composition? *Photosynth. Res.*, **40**, 107.

RUTHERFORD, A. W. and ZIMMERMAN, J. L. (1984) A new EPR signal attributed to the primary plastoquinone acceptor of photosystem II, *Biochim. Biophys. Acta*, **767**, 168.

SANDMANN, G. and MALKIN, R. (1984) Light inhibition of respiration is due to a dual function of cytochrome b_6/f complex and the PC/cyt c_{553} pool in *Aphanocapsa*, *Arch. Biochem. Biophys.*, **234**, 105.

SATOH, K. and FORK, D. C. (1983) The relationship between state II to state I transitions and cyclic electron flow around photosystem I, *Photosynth. Res.*, **4**, 245.

SCHATZ, G. H. and VAN GORKOM, H. J. (1985) Absorbance difference spectra upon charge transfer to secondary donors and acceptors in photosystem II, *Biochim. Biophys. Acta*, **810**, 283.

SCHELLER, H. V., SVENDSEN, I. and LINDBERG-MOLLER, B. (1989) Subunit comparison of photosystem I and identification of center X as a 4Fe–4S iron sulphur center, *J. Biol. Chem.*, **264**, 6929.

SCHERER, S. (1990) Do photosynthetic and respiratory electron transport chains share redox components? *Trends Biochem. Sci.*, **15**, 458.

SCHREIBER, U., BILGER, W. and SCHLIWA, U. (1986) Continuous recording of photochemical and non-photochemical quenching with a new type of modulation fluorometer, *Photosynth. Res.*, **10**, 51.

SHAHAK, Y., ARIELI, B., BINDER, B. and PADAN, E. (1987) Sulphide dependent photosynthetic electron flow coupled to proton translocation in thylakoids of cyanobacterium *Oscillatoria limnetica*, *Arch. Biochem. Biophys.*, **259**, 605.

SHUBIN, V. V., BEZMERTANYA, I. N. and KARAPETYAN, N. V. (1992) Isolation from

Spirulina membranes of two photosystem I type complexes, one of which contains the chlorophyll responsible for 77 K fluorescence band at 760 nm, *FEBS Lett.*, **309**, 340.

SHUBIN, V. V., MURTHY, S. D. S., KARAPETYAN, N. V. and MOHANTY, P. (1991) Origin of the 77 K variable fluorescence at 758 nm in the cyanobacterium *Spirulina platensis*, *Biochim. Biophys. Acta*, **1060**, 28.

SHUBIN, V. V., TSUPRUN, V. L., BEZMERTANYA, I. N. and KARAPETYAN, N. V. (1993) Trimeric forms of the photosystem I reaction center complex preexist in the membranes of cyanobacterium *Spirulina platensis*, *FEBS Lett.*, **334**, 79.

SHUVALOV, V. A., NUIJS, A. M., VAN GORKOM, H. J., SMITH, H. W. J. and DUYSENS, L. M. N. (1986) Picosecond absorbance changes upon selective excitation of the electron donor P_{700} in photosystem I, *Biochim. Biophys. Acta*, **850**, 319.

SHYAM, R. and SANE, P. V. (1989) Photoinhibition of photosynthesis and its recovery in low and high light acclimatized blue-green alga (cyanobacterium) *Spirulina platensis*, *Biochem. Physiol. Pflanzen*, **185**, 211.

SIVAK, M. N. and VONSHAK, A. (1988) Photosynthetic characteristics of *Spirulina platensis* on solid support. Chlorophyll fluorescence kinetics, *New Phytol.*, **110**, 241.

TANDEAU DE MARSAC, N. (1991) Chromatic adaptation by cyanobacteria. In BOGORAD, L. and VASIL, I. K. (Eds) *The Photosynthetic Apparatus – Molecular Biology and Operation*, p. 417, San Diego: Academic Press.

TANG, X. S. and SATOH, K. (1984) Characterization of a 47 Kd chlorophyll binding polypeptide (CP47) isolated from Photosystem II core complex, *Plant Cell Physiol.*, **25**, 935.

TISCHER, W. and STROTMANN, H. (1977) Relationship between inhibitor binding by chloroplasts and inhibition of photosynthetic electron transport, *Biochim. Biophys. Acta*, **460**, 113.

TREBST, A. (1986) Topology of the plastoquinone and herbicide binding peptides of photosystem II in thylakoid membrane., *Z. Naturforsch.*, **41C**, 240.

VAN GRONDELLE, R. and AMESZ, J. (1986) Excitation energy transfer in photosynthetic systems. In GOVINDJEE, AMESZ, J. and FORK, D. C. (Eds) *Light Emission by Plants and Bacteria*, p. 191, London: Academic Press.

VERMAAS, W. F. J. and IKEUCHI, M. (1991) Photosystem II. In BOGORAD, L. and VASIL, I. K. (Eds) *The Photosynthetic Apparatus – Molecular Biology and Operation*, p. 25, San Diego: Academic Press.

VERMAAS, W. F. J., IKEUCHI, M. and INOUE, Y. (1988a) Protein composition of the photosystem II core complex in genetically engineered mutants of the cyanobacterium *Synechocystis* PCC 6803, *Photosynth. Res.*, **17**, 97.

VERMAAS, W. F. J., RUTHERFORD, A. W. and HANSSON, O. (1988b), Site directed mutagenesis in photosystem II of the cyanobacterium *Synechocystis* PCC 6803: Donor D is a tyrosine residue in the D2 protein, *Proc. Natl. Acad. Sci.* (USA), **85**, 8477.

VONSHAK, A., GUY, R., POPLAWSKY, R. and OHAD, I. (1988) Photoinhibition and its recovery in two different strains of *Spirulina*, *Plant Cell Physiol.*, **29**, 721.

WHITMARSH, J. (1986) Mobile electron carriers in thylakoids. In STAEHELIN, L. A. and ARNTZEN, C. J. (Eds) *Encyclopaedia of Plant Physiology – New Series*, Vol. 19, p. 508, Berlin: Springer-Verlag.

WILD, A., HOPFNER, M., RUHLE, W. and RICHTER, M. (1986) Changes in the stoichiometry of photosystem II components as an adaptive response to high light and low light conditions during growth, *Z. Naturforsch.*, **41C**, 597.

WILHELM, C. and WILD, A. (1984) The variability of the photosynthetic unit in *Chlorella*, *J. Plant Physiol.*, **115**, 125.

YAMAGUCHI, N., TAKAHASHI, Y. and SATOH, K. (1988) Isolation and characterization of a photosystem II core complex depleted in the 43 Kd chlorophyll binding subunit, *Plant Cell Physiol.*, **29**, 123.

3

Spirulina: Growth, Physiology and Biochemistry

AVIGAD VONSHAK

Introduction

Algal physiology and biochemistry have been reviewed and discussed quite extensively in the last decades. The excellent contributions by Lewin (1962), Fogg (1975) and Carr and Whitton (1973) are just a few examples of textbooks which cover a wide range of topics related to algal physiology and biochemistry. The aim of this chapter is to point out relevant areas in which *Spirulina* has been used as a model organism or studies whose data can be of significant importance in further understanding the growth, physiology and biochemistry of *Spirulina*, especially when grown in outdoor conditions.

Growth Rate: The Basics

The growth rate of *Spirulina* follows the common pattern of many other microorganisms which undergo a simple cell division without any sexual or differentiation step. Thus, under 'normal' growth conditions the specific growth rate (μ) is described by the following equation

$$\mu = \frac{t}{x} \frac{dx}{dt} \tag{3.1}$$

where x is the initial biomass concentration. The way to calculate the specific growth rate of microalgae has been described in many publications (Vonshak, 1986, 1991; Stein, 1973). The most commonly used formula is:

$$\mu = \frac{\ln x_2 - \ln x_1}{t_2 - t_1} \tag{3.2}$$

where x_1 and x_2 are biomass concentrations at time intervals t_1 and t_2. The simple equation that combines the specific growth rate (μ) and the doubling time (d.t.) or the generation time (g) of a culture is:

$$g = \frac{\ln 2}{\mu} = \frac{0.693}{\mu} = \text{d.t.} \tag{3.3}$$

These equations are true for the logarithmic or exponential phase of growth in batch cultures. When growing an algal culture in a continuous mode such as in a chemostat or turbidostat the equations are modified so that

$$\mu = D = \frac{1}{v} = \frac{dv}{dt} \tag{3.4}$$

where v is the total volume of the culture and dv/dt is the dilution rate.

More detailed studies on growth kinetics of *Spirulina* were performed by Lee et al. (1987) and Cornet et al. (1992a, b). The studies included elaborate in-depth details of mathematical modeling, which are beyond the scope of this chapter.

Growth Yield and Efficiency of Photoautotrophic Cultures

Many of the studies on *Spirulina* that attempted to estimate its growth yield and photosynthetic efficiency were limited, mainly because most of the cultures were not axenic (bacteria-free). Developing procedures to obtain an axenic culture of *Spirulina* led the way to this kind of study (Ogawa and Terui, 1970). The first assessment of quantum yield for *Spirulina* using cultures grown at different dilution rates was in a Roux bottle. The opalescent plate method was used to measure the light energy absorbed by the cells and to assess the growth yield, Y_{kcal}, i.e. the amount of dry algal biomass harvested per kcal light energy absorbed. Calculated values of Y_{kcal} ranged from 0.01 to 0.02 g cell kcal^{-1}. These values corresponded to a Y value of 6–12 per cent. In a much later study (Ogawa and Aiba, 1978) where assimilation of CO_2 was used to estimate the quantum requirement of *Spirulina* cultures grown at steady state conditions, it was found that the Q_{CO_2} value was about 20 quantum mol^{-1} CO_2, which corresponds to a Y value of 10 per cent. This is in good agreement with the Y_{kcal} values of 0.01–0.02 reported earlier.

The relation of the specific growth rate to the specific absorption rate of light energy was used to establish a mathematical equation describing the growth of *Spirulina* in a batch culture (Iehana, 1987). The equation indicates that the specific growth rate increases linearly with the increase of the specific absorption rate of light energy in culture with a high cell concentration. In an earlier work, Iehana (1983) analyzed the growth kinetics of *Spirulina* when grown as a continuous culture under light limitation. The kinetic analysis was done by comparing the relationships between the extinguished luminous flux in the culture and the growth rate. Under fixed luminous conditions, the specific growth rate of *Spirulina* was proportional to the extinction rate of the luminous flux per cell concentration. The obtained equation simulated growth in the exponential phase. When cell concentration was kept constant, the equation was comparable to Michaelis–Menten type kinetics. Two other groups, Lee et al. (1987) and Cornet et al. (1992a, b) have published detailed studies on attempts to establish a mathematical model for the growth of *Spirulina* under a variety of growth conditions. It seems that they all fit well the experimental growth data under normal steady state conditions, where light is either limiting or is at its saturation level. These models should be modified if stress conditions, such as photoinhibition or environmental stress (i.e. temperature or salinity), are introduced.

Mixotrophic and Heterotrophic Growth

The isolation of an axenic culture of *Spirulina* enabled the use of different organic sources to stimulate growth, in heterotrophic or mixotrophic modes. Ogawa and Terui (1970) were the first to report that an axenic culture of *Spirulina* grown on a mineral medium enriched with 1 per cent peptone had a higher growth rate in the logarithmic phase and in the linear phase than the culture grown on minimal medium. There was a 1.2–1.3 fold and 1.85–1.93 fold increase, respectively.

The effectiveness of the peptone was higher in light-limited cultures. Addition of glucose also affected the growth yield and a combination of 0.1 per cent peptone and 0.1 per cent glucose doubled growth yield compared with that obtained without any organic carbon source. Cultures of *Spirulina* grown on glucose were used to further analyze the autotrophic and heterotrophic characteristics of the cells.

Stimulation of the growth rate in the presence of glucose suggests that respiratory activity occurs in *S. platensis* even in light. Photosynthetic (O_2 evolution) and respiratory (O_2 consumption) activities were examined using 100-h and 250-h cultures grown on glucose, either in the light or aerobic-dark conditions in the presence or absence of 5 mM DCMU (3-(3,4-dichlorophenyl)-1,dimethylurea, a potent inhibitor of photosynthesis). Respiratory activity clearly indicated that the rate of O_2 consumption was unaffected by light, irrespective of DCMU presence. In the presence of DCMU no photosynthetic activity was detected. Heterotrophically grown cells also showed lower photosynthetic activity. This might be due to the fact that the contents of pigments such as chlorophyll-a , carotenoids and phycocyanin in the cells were lower than in the autotrophic and mixotrophic cultures. Results indicated that in mixotrophic conditions, autotrophic and heterotrophic growth functions independently in *S. platensis* without interaction (Marquez et al., 1993). In a recent publication (Marquez et al., 1995), the potential of heterotrophic, autotrophic and mixotrophic growth of *Spirulina* was evaluated. Most of the results agree with those previously published, except that in this case, heterotrophic growth of *Spirulina* was observed in cultures grown in the presence of glucose. Perhaps in previous studies, the cultivation time was not long enough. Marquez et al. (1995) also suggest that CO_2 produced from heterotrophic glucose metabolism might be used photosynthetically, together with bicarbonate from the culture medium.

These results have to be further investigated in other *Spirulina* strains so that it can be established whether these characters are universal for most of the *Spirulina* strains or specific to the strain used by the researcher.

Response to Environmental Factors

The Effect of Light

Without doubt, light is the most important factor affecting photosynthetic organisms. Due to the prokaryotic nature of *Spirulina*, light does not affect the differentiation or development processes. Nevertheless, *Spirulina*, like many other algae grown photoautotrophically, depends on light as its main energy source. The photosynthetic apparatus and its components are described in Chapter 2.

The response of outdoor cultures to light and the important role that light and photosynthesis play in productivity in mass cultivation of *Spirulina* are discussed in detail in Chapters 5 and 8. In this section, we will examine the effect of light on laboratory cultures of *Spirulina*, and the way cells respond and adapt to different levels of light.

Effect on growth

Most of the laboratory studies on the response of *Spirulina* to light were performed under photoautotrophic growth conditions, using a mineral medium and bicarbonate as the only carbon source. The first detailed study on the response of *Spirulina maxima* to light was done by Zarrouk in 1966. In his somewhat simple experiment, he reached the conclusion that growth of *S. maxima* is saturated at levels of 25–30 klux. Since not much information is given on the way light was measured and the light path in the vessel cultures, it is very difficult to compare these results with more recent ones. From data obtained in the author's laboratory, growth of *Spirulina platensis* became saturated at a range of 150–200 μmol m^{-2} s^{-1}. This is about 10 to 15 per cent of the total solar radiance at the 400–700 nm range. This value is highly dependent on growth conditions and correlates with the chlorophyll to biomass concentration. Another experimental parameter which determines this response is the light path of the culture. Therefore, it is highly recommended that when attempting to establish the maximal specific growth rate μ_{max}, a turbidostat system should be employed. In such a manner, we have estimated the μ_{max} of *Spirulina* to be in the range of 8–10 h. The use of a turbidostat system also eliminates nutritional limitation or self-shading problems.

Effect on photosynthesis

The most common way to study the photosynthetic response of algal cultures to light is through the measurement of the photosynthesis (P) versus irradiance (I) curves. A typical $P–I$ curve is shown in Figure 3.1. The saturation and compensation points are

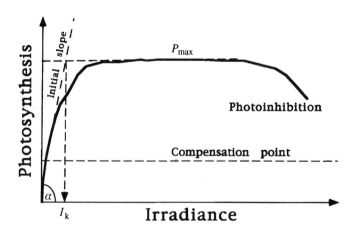

Figure 3.1 Schematic diagram of a photosynthesis (P) versus irradiance curve, showing the typical photosynthetic parameters. For more details see text.

the most important parameters. In the dark, the rate of oxygen evolution or carbon fixation will be negative because of respiration. As irradiance is increased, a point is reached when the photosynthetic rate is just balanced by respiration. This is the compensation point. As irradiance is further increased, the rate of photosynthesis increases linearly. Eventually, the curve levels off, as photosynthesis becomes saturated, reaching a maximum, P_{max}. The initial slope, α, is a useful indicator of quantum yield, i.e. photosynthetic efficiency.

The saturation irradiance may be also defined by the value of I_k, which represents the point at which the extrapolation of the initial slope crosses P_{max}. Exposing *Spirulina* cultures to high photon flux densities above the saturation point may result in a reduction of the rate of photosynthesis, a phenomenon defined as photoinhibition. The classical view that photoinhibition is observed only at high irradiance values today seems to be a very simplistic one. It will be discussed later how photoinhibition may be observed even at relatively low irradiance levels when other environmental stresses are introduced.

The P_{max} and I_k levels are highly dependent on growth conditions. *Spirulina* cultures grown at high or low light intensities will have different P_{max} and I_k values. Changes in these values may represent the culture's ability to photoadapt to the different light environments. Furthermore, the P_{max} and I_k values may be used as a tool for screening strains of *Spirulina* which have a better photosynthetic performance under outdoor conditions. An example for such a screening process is given in Table 3.1, a summary of experiments carried out in the author's lab, indicating different photosynthetic parameters in three different *Spirulina* strains. Although the strains were grown under the same temperature and light conditions, they have different α and I_k values. The fact that the cultures have a similar growth rate, μ, under laboratory conditions may be meaningless for the outdoor cultivation systems. For the outdoor conditions, strains with different P_{max} or I_k may have different productivities since they differ in their ability to utilize the high solar irradiance available outdoors.

Table 3.1 Photosynthetic parameters obtained from *P* versus *I* curves of three *Spirulina* isolates, grown under the same laboratory conditions

	Spirulina strain		
Parameters	BP	P4P	Z19/2
μ	0.048	0.043	0.044
I_k	145 ± 15	115 ± 13	165 ± 15
P_{max}	625 ± 8	614 ± 5	645 ± 13
α	4.8 ± 0.4	6.3 ± 1.0	3.85 ± 0.7

μ is specific growth rate (h^{-1}).
I_k is irradiance at the onset of light saturation ($\mu E\,m^{-2}\,s^{-1}$).
P_{max} is maximal rate of light-saturated photosynthesis.
α is initial slope of the *P–I* curve ($\mu mol\ O_2\ h^{-1}\,mg\ Chl^{-1}$)/ ($\mu E\,m^{-2}\,s^{-1}$).
All strains were grown under laboratory conditions, constant temperature of 35 °C and constant light, 120 $\mu mol\,m^{-2}\,s^{-1}$.

Light stress – photoinhibition

Photoinhibition, as mentioned earlier, is defined as a loss of photosynthetic capacity due to damage caused by photon flux densities (PFD) in excess of that required to saturate photosynthesis. The phenomenon of photoinhibition has been studied extensively and is well documented in algae and higher plants (Critchley, 1981; Greer et al., 1986; Kyle and Ohad, 1986; Öquist, 1987; Powles, 1984).

The phenomenon of photoinhibition in laboratory *Spirulina* cultures was first studied by Kaplan (1981), who observed a reduction in the photosynthetic activity when cells of *Spirulina* were exposed to high light under CO_2-depleted conditions. It was suggested that the reduction of the photosynthetic activity was due to the accumulation of H_2O_2.

A much more elaborate study on the photoinhibitory response was carried out in our laboratory (Vonshak et al., 1988a). We demonstrated that different strains of *Spirulina* may differ in their sensitivity to the light stress. At least in one case it was found that this difference was most likely due to a different rate of turnover of a specific protein, D1, which is part of the PS II (see Chapter 2). The different response of *Spirulina* strains to a photoinhibitory stress may be a genotypic characteristic as well as arising from growth conditions. We also found that cultures grown at high light intensity exhibit a higher resistance to photoinhibition, as demonstrated in Figure 3.2. Cultures grown at 120 and 200 $\mu mol\, m^{-2}\, s^{-1}$ were exposed to a HPFD of 1500 $\mu mol\, m^{-2}\, s^{-1}$. Indeed, the cells grown in strong light do show a higher resistance to light stress. It should be emphasized that the photoinhibition not only affects the P_{max} level, but actually has a stronger effect on light-limited photosynthetic activity. This can be shown by comparison of the $P-I$ curves of the control and the photoinhibited cultures of *Spirulina* (Figure 3.3). It can be seen that photoinhibited cultures have a lower photosynthetic efficiency. Therefore, they are more light-limited than the control cultures, i.e. requiring more light in order to achieve the same photosynthetic activity. The implications for outdoor cultures of this observation will be discussed in Chapter 5.

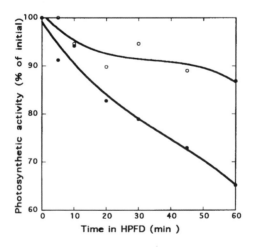

Figure 3.2 The effect of growth irradiance on the response of *Spirulina* to HPFD. Cells were grown at 120 (●) or 200 (o) $\mu mol\, m^{-2}\, s^{-1}$ for 4 days and then diluted to the same chlorophyll concentration and exposed to HPFD of 1500 $\mu mol\, m^{-2}\, s^{-1}$.

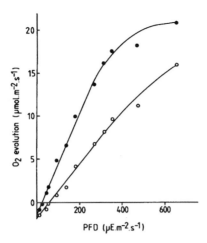

Figure 3.3 Photosynthesis versus irradiance curves of control (●) and photoinhibited (o) *Spirulina* cultures.

Effect of Temperature

While light is considered the most important environmental factor for photosynthetic organisms, temperature is undoubtedly the most fundamental factor for all living organisms. Temperature affects all metabolic activities. Temperature also affects nutrient availability and uptake, as well as other physical properties of the cells' aqueous environment.

Effect of temperature on growth

Spirulina was originally isolated from temporal water bodies with a relatively high temperature. The usual optimal temperature for laboratory cultivation of *Spirulina* is

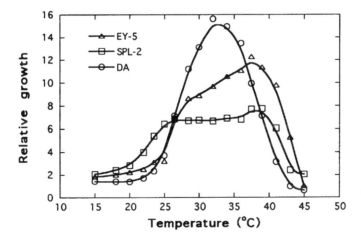

Figure 3.4 The response of three *Spirulina* isolates to temperature, as measured by the increase in chlorophyll concentration in cultures incubated in a temperate gradient block of 15–45 °C illuminated continuously at 150 $\mu mol\, m^{-2} s^{-1}$.

in the range 35–38 °C. However, it must be pointed out that this range of temperature is arbitrary. Many *Spirulina* strains will differ in their optimal growth temperature, as well as their sensitivity to extreme ranges. In our laboratory, many strains of *Spirulina* are maintained and tested for their physiological responses. In Figure 3.4, three isolates of *Spirulina* are compared. The cells were incubated under constant light in a temperature gradient block. The increase in chlorophyll was measured after a certain period of incubation. The three strains differed significantly in their response to temperature. The one marked DA has a relatively low temperature optimum of 30–32 °C, while the one marked EY-5 grows well at temperature of up to 40–42 °C. The isolate marked SPL-2 is characterized by a relatively wide temperature optimum. This is just one example of the variations observed. Obtaining strains with a wide temperature optimum could be of high monetary value in the outdoor cultivation industry since we believe that temperature is one of the most important limiting factors in outdoor production of *Spirulina*. Specific strains which fit the local climatic conditions should be used.

Effect of temperature on photosynthesis and respiration

The net productivity of an algal culture is directly correlated to the gross rate of CO_2 fixation or O_2 evolution (photosynthesis) and the rate of respiration. Photosynthesis and respiration are dependent on temperature, but only CO_2 fixation and O_2 evolution are both light- and temperature-dependent. A detailed study on the response of a *Spirulina* strain marked M-2 was performed by Torzillo and Vonshak (1994). The O_2 evolution rate of *Spirulina* cells measured at different temperatures is shown in Figure 3.5. The optimal temperature for photosynthesis was 35 °C; however, growth at 28 per cent and 23 per cent of the optimum were measured at the extreme minimum and maximum temperatures tested: 10 °C and 50 °C, respectively. The effect of temperature on the dark respiration rate of *Spirulina* was also measured, by following the O_2 uptake rate in the dark. A temperature-dependent exponential relationship was obtained, with the respiration rate increasing as temperature increased (Figure 3.6).

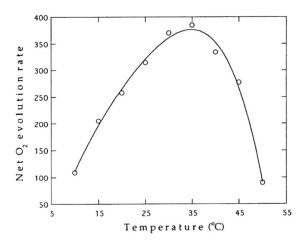

Figure 3.5 The effect of temperature on the gross O_2 evolution rate (photosynthesis) (μmol O_2 mg chl^{-1} h^{-1}) of *Spirulina platensis* cells. Cells were allowed to equilibrate at each temperature for 15 min before the measurement.

The temperature-dependent dark respiration rate was given by

$$R = 0.771e^{(0.06367T)}$$

where R is the respiration rate (μmol O_2 mg^{-1} chl h^{-1}) and T is the temperature (°C). At 50 °C and 15 °C dark respiration rates dropped almost to zero. An Arrhenius plot for respiration showed an activation energy of 48.8 kJ mol^{-1} for *Spirulina*. The temperature coefficient (Q_{10}) of the organism in a temperature range was calculated by the following equation, deduced from the Arrhenius equation (Pirt, 1975):

$$\log Q_e = \frac{E_a}{2.303R} \frac{10}{(T+10)T} \tag{3.5}$$

where E_a is the activation energy (kJ mol^{-1}) and R is the universal gas constant (8.31 J K^{-1} mol^{-1}). A Q_{10} of 1.85 was calculated for the range 20–40 °C. The respiration-to-photosynthesis ratio in *Spirulina* was 1 per cent at 20 °C and 4.6 per cent at 45 °C. These rather low values confirm the general assumption that cyanobacteria have low respiration rates (van Liere and Mur, 1979). The respiration-to-photosynthesis rates measured in these experiments were found to be much lower than those reported for outdoor cultures of *Spirulina*, where up to 34 per cent of the biomass produced during the daylight period may be lost through respiration at night (Guterman et al., 1989; Torzillo et al., 1991). However, it must be noted that respiration rate is strongly influenced by light conditions during growth. In our *Spirulina* strain, the respiratory activity had a much higher temperature optimum than the photosynthetic activity. Nevertheless, the photosynthetic activity of the cells was more resistant to the temperature extremes than dark respiration at the minimum and maximum temperatures tested.

Interaction of temperature and light

Deviation from the optimal growth temperature has an inhibitory effect on the photosynthetic capacity. This reduction in activity represents a limitation that is

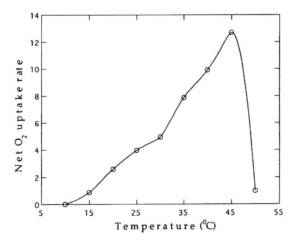

Figure 3.6 The effect of temperature on O_2 uptake rate in the dark (respiration) (μmol O_2 mg chl^{-1} h^{-1}) of *Spirulina platensis* cells. Cells were allowed to equilibrate at each temperature for 15 min before the measurement.

immediately overcome after a shift back to the optimal temperature, if no other damage was done. The kinetics of recovery from low-temperature incubation indicate that some repair mechanism must take place before the original photosynthetic activity is reached. This observation was made only when the cultures were incubated at a low temperature in the light. *Spirulina* cultures incubated at a low temperature in the dark seemed to acquire their original photosynthetic activity as soon as they were transferred to 35 °C without any lag period. It is thus suggested that *Spirulina* cultures grown at less than the optimal temperature are more sensitive to photoinhibition than those grown at the optimal temperature. The latter will be better able to handle excess light energy, since they have a higher rate of electron transport, an active repair mechanism and more efficient ways of energy dissipation. As shown in Figure 3.7, cultures exposed to HPFD at 25 °C, a temperature below the optimal, were much more sensitive to HPFD stress, as compared with cultures exposed to the HPFD at 35 °C. The difference was more pronounced with prolonged exposure time to high irradiance. This fits well with the overall concept of photoinhibition, i.e. that any environmental factor which reduces the rate of photosynthesis may encourage photoinhibition. Jensen and Knutsen (1993) have demonstrated that the increased susceptibility of *Spirulina* to HPFD at low temperatures may also be due to a lower rate of protein synthesis, affecting cells' recovery from light stress.

Many other factors interact with temperature and probably affect the growth and productivity of *Spirulina*. Solubility of gases in the medium and availability of nutrients are some of these. More detailed and extensive work is required in order to better understand these interactions.

Response to Salinity

Cyanobacteria inhabit environments which vary drastically in their saline levels. In the last 15 years many studies were published on the response of cyanobacteria to

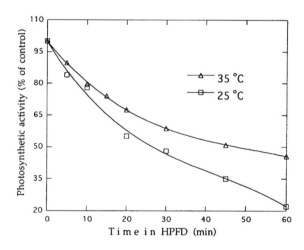

Figure 3.7 The effect of temperature on the response of *Spirulina* to a HPFD stress. Cultures incubated at 35 °C or 25 °C were exposed to 2500 μmol m^{-2}s^{-1}. At time intervals, the reduction in photosynthetic activity was measured.

different saline environments: the specific role of organic compounds as osmoregulants (Borowitzka, 1986), modification in photosynthesis and respiration activity (Vonshak and Richmond, 1981), and variations in the protein synthesis pattern (Hagemann et al., 1991). Different *Spirulina* species have been isolated from a variety of saline environments. We will describe the work done using strains isolated from alkaline and brackish water. Since the exact taxonomic position of the marine strains of *Spirulina* is still not clear, we will not discuss their response, although it has been a subject of a detailed study (Gabbay and Tel-Or, 1985).

Effect of salinity on growth

Exposure of *Spirulina* cultures to high NaCl concentrations results in an immediate cessation of growth. After a lag period, a new steady state is established. A typical growth response curve to NaCl is shown in Figure 3.8, where changes in biomass concentrations of three *Spirulina* cultures exposed to control, 0.5 and 0.75 M NaCl are presented. As can be seen, not only is growth inhibited for at least 24 h after the exposure at the high NaCl concentration, but a decrease in biomass is observed after which a new steady-state exponential growth rate is established. The new growth rates after adaptation are slower and inversely correlated to the increased NaCl concentration in the medium (Vonshak et al., 1988b). A decrease in the growth rate because of salt stress has also been demonstrated in other cyanobacteria, such as *Anacystis* (Vonshak and Richmond, 1981) and *Nostoc* (Blumwald and Tel-Or, 1982). It is worth noting that the length of the time lag is exponentially correlated to the degree of stress imposed on the cells. This lag period in many cases is associated with a decline in chlorophyll and biomass concentrations in the culture (Vonshak et al., 1988b).

The response of *Spirulina* to salinity with regard to degree of growth inhibition, adaptability to salt levels and the rate of adaptation varies widely, depending on the strain used in the study. In Table 3.2, an example is given for two strains of *Spirulina* exposed to different salt concentrations. The changes in growth rate and doubling time after adaptation indicate that the M2 strain seems to be more resistant to the salt stress than the 6MX strain.

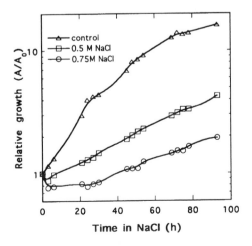

Figure 3.8 The growth response of *Spirulina* to increased concentrations of NaCl in the growth medium (NaCl concentrations indicated are above the normal level in the growth medium).

Table 3.2 Specific growth rates and doubling time of *Spirulina* strains grown under salinity stress at 35 °C

Treatment	M2		6MX	
	Specific growth rate (h^{-1})	Doubling time (h)	Specific growth rate (h^{-1})	Doubling time (h)
Control	0.063	11.0	0.059	11.8
+0.50 M NaCl	0.044	15.9	0.026	26.2
+0.75 M NaCl	0.034	20.3	0.018	39.4

Effect of salinity on photosynthesis and respiration

It has been suggested that exposure to high salinity is accompanied by a higher demand for energy by the stressed cells (Blumwald and Tel-Or, 1982). Changes in the photosynthetic and respiratory activity of *Spirulina* were measured over a period of 30 min to 48 h, after exposure to 0.5 and 1.0 M NaCl. These changes were compared with changes in biomass concentration as an indicator of growth (Figure 3.9). A marked decrease in the photosynthetic oxygen evolution rate was observed 30 min after exposure to the salt at both concentrations (Figure 3.9a). This decline was followed by a recovery period, characterized by a lower steady-state rate of photosynthesis. Recovery at 0.5 M NaCl was faster than at 1 M NaCl (after 1.5 vs 3.0 h) and leveled off at 80 per cent of the control activity vs about 50 per cent respectively. Respiratory activity also dropped rapidly immediately after salt application at both concentrations (Figure 3.9b). Activity was restored ten times faster at 0.5 M than at 1.0 M NaCl and continued to increase to twice the control level at 1.0 M NaCl.

The immediate inhibition of the photosynthetic and respiratory systems after exposure to salt stress was explained by Ehrenfeld and Cousin (1984) and Reed et al. (1985). They showed that a short-term increase in the cellular sodium concentration was due to a transient increase in the permeability of the plasma membrane during the first seconds of exposure to high salt. It has been suggested that the inhibition of photosynthesis arising from the rapid entry of sodium, might be the result of the detachment of phycobilisomes from the thylakoid membranes (Blumwald et al., 1984). Elevated activities of dark respiration in cyanobacteria because of salinity stress have previously been reported (Vonshak and Richmond, 1981; Fry et al., 1986; Molitor et al., 1986). This high activity may be associated with the increased level of maintenance energy required for pumping out the toxic sodium ions.

Osmoregulation and strain-specific response of Spirulina to salinity

During the course of adaptation to salinity, an osmotic adjustment is required. In *Spirulina*, a low molecular weight carbohydrate accumulates. This has been identified as a nine-carbon heteraside named Glucosyl-glycerol, as well as trealase (Martel et al., 1992). We compared biomass composition of two *Spirulina* strains grown under salt stress conditions; a significant change in biomass composition was observed, mainly reflected in the increase in carbohydrates and a decrease in the

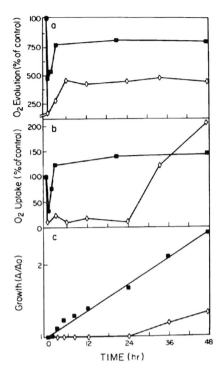

Figure 3.9 Effect of NaCl on photosynthesis (a), respiration (b) and growth (c) in *Spirulina* exposed to ■—■ 0.5 M and ◇—◇ 1.0 M NaCl. 100 per cent activity values for apparent photosynthesis and respiration were 663 and 70 μmol O_2 mg chl^{-1}h^{-1}, respectively

protein level (Table 3.3). These changes are correlated with the degree of stress imposed, i.e., a higher level of carbohydrates at a higher salt concentration. The difference in the level of carbohydrates accumulated by the two strains may also reflect a difference in their ability to adapt to salt stress.

Interaction with light

Photosynthetic activity of *Spirulina* declines under salinity stress even if the cultures are grown continuously in the saline environment and adapt to the new osmoticum. This decline is associated with a modification in the light energy requirement, i.e.

Table 3.3 Biomass composition of *Spirulina* strains grown for 95 h under salinity stress

Strain	Treatment	Chlorophyll (% dry weight)	Protein (% dry weight)	Carbohydrate (% dry weight)
M2	Control	2.6	56.1	35.5
	+0.50 M NaCl	1.9	44.1	46.7
	+0.75 M NaCl	1.3	41.1	55.8
6MX	Control	1.6	60.2	30.0
	+0.50 M NaCl	1.0	23.9	64.4
	+0.75 M NaCl	0.5	22.0	61.4

Table 3.4 Photosynthetic characteristics of *Spirulina* grown under control and salinity conditions

	Control	0.68 M Nacl
μ	0.048	0.024
I_k	160	250
P_{max}	625	285
α	3.9	1.1
R	24	63

μ = specific growth rate (h^{-1}).
I_k = light saturation ($\mu E\,m^{-2}\,s^{-1}$).
P_{max} = saturated rate of photosynthesis ($\mu mol\ O_2\ h^{-1}\mu g\ Chl^{-1}$).
R = dark respiration ($\mu mol\ O_2$ uptake $h^{-1}\mu g\ Chl^{-1}$).
α = initial slope of the P–I curve ($\mu mol\ O_2\ h^{-1}mg\ Chl^{-1}$)/($\mu E\,m^{-2}\,s^{-1}$).

less light is required for the saturation of photosynthesis. Comparison of the P vs I curve for control and salt stress cultures shows that all the parameters have been modified (Table 3.4), indicating a reduction in the P_{max} as well as the photosynthetic efficiency, known as the initial slope α. The amount of light required to saturate photosynthesis (I_k) increased, again indicating that photosynthesis operates at a significantly lower efficiency when *Spirulina* is exposed to salt stress. It is also worth noting that the dark respiration rate is increased by 2.5 fold, which may be how cells produce the extra energy required to maintain their internal osmoticum. This reduction in the ability to use light energy absorbed by the photosynthetic apparatus increases the sensitivity of the salt-stressed cells to photoinhibition. When salt-stressed cultures of *Spirulina* are exposed to HPFD, a much faster decline in photosynthesis is observed, as compared with the control (Figure 3.10). Control cultures exposed to a photoinhibitory stress lose about 40 per cent of their photosynthetic activity after 60 min exposure, in 0.5 M and 0.75 M NaCl cultures: a

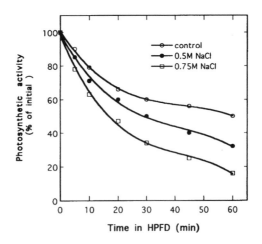

Figure 3.10 The response of *Spirulina* cells grown at different NaCl concentrations to a photoinhibitory stress. Cells were allowed to adapt to the salinity stress and only then exposed to a HPFD of 2500 $\mu mol\,m^{-2}\,s^{-1}$

60 per cent and 80 per cent reduction is observed, respectively. Most likely, salinity-stressed cells are less efficient in handling light energy (Table 3.4) and also have a lower rate of protein synthesis. Since recovery from photoinhibition is associated with the ability to synthesize specific protein associated with PS II (Vonshak et al. 1988a), a reduction in the level of the protein synthesis affects the repair mechanism.

Conclusions

Although it has been almost more than 20 years since the commercial application of *Spirulina* was proposed, relatively very little has been done to study the basic physiology of *Spirulina*. The original work of Zarrouk (a Ph.D. thesis written in France in 1966) was never published in a scientific journal, which was most unfortunate since it contained virtually unknown, valuable information about *Spirulina*. The unclear situation of the systematics of *Spirulina* (Chapter 1) has made comparative physiological studies even more difficult. We have tried to summarize most of the recent findings on *Spirulina* growth physiology; we believe that further in-depth study to identify *Spirulina* strains and measure their response to environmental factors is required. We also believe that with the new molecular biology studies of *Spirulina*, more information and a better understanding will be achieved.

Biochemistry

Introduction

The use of *Spirulina* as an experimental tool in biochemical studies has been very limited. The number of publications related to metabolic pathways and enzyme isolation is low. This dearth of information shows that the main interest in *Spirulina* is for its biotechnological application. *Spirulina* does not fix nitrogen and does not develop differentiated cells like heterocysts or akinates as part of the filament. Since most of the isolates do not form colonies when grown on a solid support, genetic manipulation of *Spirulina* is difficult. Moreover, only a few laboratories have reported the isolation of axenic (bacteria-free) cultures.

Elongation Factor

EF-Tu, the elongation factor that binds aminoacyl-tRNA to the ribosome, plays an important role in the biosynthesis of proteins. It has been purified from a number of bacteria, as well as from chloroplast of higher plants and green algae.

The EF-Tu of *S. platensis* was first isolated by Tiboni and Ciferri (1983). It appeared to be very similar to the protein isolated from bacteria. The estimated molecular weight of *S. platensis* EF-Tu is about 50 000, similar to that reported for the EF-Tu of Gram-positive bacteria. The protein was first isolated using a Sephadex G-100 column, and an EF-Tu-containing fraction was identified by assaying GDP binding activity. Further purification was performed by an affinity chromatography step using GDP sepharase.

EF-Tu may be used in evolutionary studies to explore phylogenetic relations between *Spirulina* and other prokaryotic groups. Attempts made by the isolators to ascertain immunological similarity between *S. platensis* EF-Tu and the protein from bacteria or chloroplasts were unsuccessful since no immunoprecipitate was observed when antisera to *E. coli* or spinach chloroplast EF-Tu were tested with crude or purified preparations of *S. platensis* EF-Tu.

Nitrite Reductase

Nitrite reduction is considered to be the final step of assimilatory nitrate reduction, in which nitrite is reduced to the level of ammonia in the following way:

$$NO_2^- + 6e^- + 8H \rightarrow NH_4^+ + 2H_2O$$

This reaction is catalyzed by the enzyme nitrite reductase (NiR).

Ferredoxin-dependent NiR (Fd-Nir) (ammonia: ferredoxin oxidoreductase, EC 1.7.7.1) has been purified from higher plants and extensively characterized. This enzyme was isolated from *Spirulina* by Yabuki et al. (1985). After breaking of the cells in a tris buffer by sonication, purification was carried out, inducing an hydrophobic chromatography anion exchange. Affinity chromatography was executed. The main steps of this process and the degree of purification are give in Table 3.5.

Yabuki et al. (1985) report that the enzyme was stable for more than a month when stored at 4 °C in a buffer containing 20 mM Tris-HCl (pH 7.5), 200 mM NaCl and 10 per cent v/v glycerol. The assay method for nitrite reductase was done in a total volume of 1 ml, 20 mmol of Tris-HCl buffer, pH 7.5, 2 mmol of sodium nitrite, 3 mmol of methyl viologen, 3.75 mg of sodium dithionite, and the enzyme preparation. The reaction was carried out at 35 °C. One unit of nitrite reductase is the amount of enzyme that reduces 1 mmol of nitrite per min under these assay conditions.

The absorption spectrum of this enzyme had six major peaks at 278, 402, 534, 572, 588 and 658 nm. This spectrum is different from that reported for spinach Fd-NiR. The nature of the visible spectrum suggests the presence of siroheme, which has been detected in enterobacterial NADPH-sulfite reductase and also in spinach sulfite reductase.

Table 3.5 Summary of the purification steps for nitrite reductase from *Spirulina*

Step	Total activity (units)	Specific activity (units [mg protein]$^{-1}$)	Purification (fold)
Crude extract	919	0.046	1
Acetone 25–75%	737	0.16	3.5
DEAE-cellulose	312	0.26	5.7
Butyl Toyopearl	216	0.40	8.7
Sephadex G-75	195	2.2	47.8
DEAE-cellulose	195	25.3	550
Fd-Sepharose 4B	78	30.0	652
Fd-Sepharose 4B	70	194	4217

Data extracted from Yabuki et al. (1985).

The molecular weight of NiR from *Spirulina* was 52 000 dalton, the same size as that of the enzyme isolated from *Anabaena*. The K_m values were 2.0×10^{-4} M (nitrite), 2.0×10^{-5} M (*Spirulina* ferredoxin), and 4.0×10^{-4} M (methyl viologen). Both prokaryotic and eukaryotic Fd-NiR had K_m values of the same order of magnitude. The pH-activity curve obtained with Tris-HCl buffer was rather broad and had an optimum at around pH 7.8.

Ferredoxin-sulfite Reductase (Fd-SiR)

The enzyme hydrogen-sulfide ferredoxin oxidoreductase, EC 1.8.7.1, which catalyzes the reduction of sulfite to sulfide with NADH, reduced ferredoxin or methyl viologen as an electron donor, was isolated from *Spirulina* by Koguchi and Tamura (1988). Their purification process yielded a highly purified enzyme, up to a homogenous band in electrophoresis test. The purification steps and the increase in purity are summarized in Table 3.6. The highly purified enzyme catalyzes the reduction of sulfite using physiological concentration of ferredoxin as an electron donor.

Comparison of the absorption spectrum of *Spirulina* Fd-SiR to that of the b-subunit (a siroheme-containing protein) of NADPH-sulfite reductase from *Escherichia coli* suggests that these enzymes are very similar in their chromophoric properties.

The molecular weight of *Spirulina* Fd-SiR obtained by native gel electrophoresis was 60 000 dalton. This value is nearly equal to those reported for MV-SiRs cited above. When tested in SDS-polyacrylamide gel electrophoresis, the purified Fd-SiR showed a molecular weight of 63 000 dalton. Koguchi and Tamura (1988) suggested that *Spirulina* Fd-SiR has a single 63 000 dalton molecular weight subunit and is composed of two identical subunits at high ionic strength.

Cytochrome b_6f

The cytochrome b_6f complex participates in electron transfer and proton translocation in photosynthesis and respiration. The b_6f complex transfers electrons

Table 3.6 Summary of the purification steps for ferredoxin-sulfite reductase

Step	Activity (units)	Specific activity (units [mg of protein] $^{-1}$)	Purification (fold)
Crude extract	720	0.066	1
Acetone (0–70%)	440	0.74	11.2
1st DEAE-cellulose	375	0.75	11.4
2nd DEAE-cellulose	238	2.5	37.9
DEAE BIO-GEL	126	11.4	172.7
Butyl-Toyopearl	83.7	22.6	342
Sephadex G-100	25.4	39.7	601
Fd-Sepharose	7.8	49.0	741

Data extracted from Koguchi and Tamura (1988).

between the two photosystems (from plastoquinol to plastocyanin), and in cyclic electron flow around photosystem I. The b_6f complex from *Spirulina* was isolated by Minami et al. (1989). The purification steps are summarized in Table 3.7.

It is worth noting that a high efficiency of recovery was obtained when hepatyl thioglucoside was used for solubilization of the thylakoid membranes. It seems that the procedure described has some advantages over the traditional sucrose gradient purification procedure.

The purified complex contained a small amount of chlorophyll and carotenoid. At least four polypeptides were present in the complex: cytochrome *f* (29 kDa), cytochrome b_6 (23 kDa), iron–sulfur protein (ISP, 23 kDa), and a 17 kDa polypeptide. Each polypeptide was separated from the complex and treated with 2-mercaptoethanol or urea. The absorption spectra of cytochrome b_6 and cytochrome *f* were similar to those of *Anabaena* and spinach, as expected. The complex was active in supporting ubiquinol-cytochrome *c* oxidoreductase activity. Fifty per cent inhibition of activity was accomplished by 1 mM dibromothymoquinone (DBMIB). The k_m values for ubiquinol-2 and cytochrome *c* (horse heart) were 5.7 mM and 7.4 mM, respectively.

The isolation of the complex and research on its structure and function may enhance understanding of the major metabolic activities in *Spirulina*, as well as providing information on the evolutionary development of photosynthesis and respiration in cyanobacteria.

ATPase activity

The ATP synthase activity of *Spirulina* was studied by several groups, most dealing with the enzyme associated with the thylakoid membrane, also known as the ATPase coupling factor, F_1.

The latent ATPase activity in photosynthetic membranes of oxygen-evolving organisms is strikingly different from other ATPase activities. Hicks and Yocum (1986) demonstrated that the latent cyanobacterial ATPase activity in *Spirulina* membrane vesicles and F_1 was elicited by treatments that stimulate chloroplast activity. They also showed that ATP acted both as an inhibitor and as an allosteric effector of CaATPase activity in *Spirulina* F_1.

Using a homogenization step for breaking the trichomes, followed by sonication, Owers-Narhi et al. (1979) obtained photosynthetic membranes from *Spirulina platensis* which contained the latent Ca^{+2}-ATPase. The purification steps used are summarized in Table 3.8.

Table 3.7 Summary of purification of the cytochrome b_6f complex from *Spirulina*

Purification step	Cyt f/protein (nmol mg^{-1})	Specific activity (µmol mg l^{-1})	Purification (fold)
Extraction	0.05	1.3	1
Solubilization	0.17	8.2	6.3
Ammonium sulfate fractionation	0.82	31.5	24.2
DEAE column chromatography	12.2	87.2	67

Data extracted from Minami et al. (1989).

Table 3.8 Partial purification of Ca^{+2}-ATPase activity from *Spirulina*

Fraction	Protein (mg)	Activity (units[a])	Sp. Activity (units mg^{-1})	Purification (fold)
Crude extract	267	16040	0.04	1.0
EDTA Supernatant	177	10610	0.12	3.0
$(NH_4)_2SO_4$ Precipitate	71	4280	0.13	3.3
DEAE Eluate	43	2600	0.651	6.1

[a] One unit is defined as 1 μmol of phosphate released per minute.

Lerma and Gomez-Lojero (1987) used *Spirulina maxima* cells in their studies. They claimed that the ATPase activity of *S. maxima* membranes did not display persistent latency as was reported for *S. platensis*. The enzyme was readily activated by similar methods used to activate the chloroplast LF_1 and showed a requirement for Mg^{2+}. The activity of ATPase reported in this study was much higher than in the one using *S. platensis* cells (Table 3.9).

Bakels et al. (1993) have recently reported in detail the unusual thermodynamic properties and activation mechanism of ATPase activity in coupled membrane vesicles isolated from *Spirulina platensis*. The nature of this activity is discussed in detail in relation to the alkalophilic nature of the cells.

Although most of the work relating to ATPase activity was done on the photosynthetic membrane and the coupling factor, it should be mentioned that ATPase activity was detected in other membrane fractions of *Spirulina*. The most recent were reported by Xu et al. (1994) describing an ATPase activity associated with the plasma membrane fraction of *Spirulina*. The activity was Mg^{+2}-dependent and could be stimulated by 50 mM of NaCl or KCl. Optimal pH reported was relatively high, 8.5, as compared with higher plant plasma membrane ATPase. This observation further supports the unique alkalophilic characteristics of *Spirulina*.

Acetohydroxy Acid Synthase (AHAS)

The enzyme acetohydroxy acid synthase (EC 4.1.3.18) is known as the first common enzyme in the biosynthesis of valine, leucine and isoleucin. It is considered to be a

Table 3.9 Purification of ATPase from *Spirulina maxima*

Fraction	Total activity (μmol Pi min^{-1})	Specific activity ([μmol Pi min^{-1}][mg protein]$^{-1}$)	Purification (fold)
Membranes	275	0.125	1.0
30–60% fraction of $(NH_4)_2SO_4$ precipitation	89.2	2.22	17.8
AF_1 after gradient centrifugation	81.43	3.62	29.0
AF_1 after ion-exchange chromatography	32.49	8.12	65.0

conserved protein, with high sequence similarities between bacteria, yeast and higher plants.

Two isoforms of acetohydroxy acid synthase were detected in cell-free extracts of *Spirulina platensis* by Forlani et al. (1991) and separated both by ion-exchange chromatography and by hydrophobic interaction. Several biochemical properties of the two putative isozymes were analyzed. It was found that they differed in pH optimum, FAD (flavin adenine dinucleotide) requirement for both activity and stability, and in heat lability. The results were partially confirmed with the characterization of the enzyme extracted from a recombinant *Escherichia coli* strain transformed with one subcloned *S. platensis*. AHAS activities, estimated by gel filtration, indicate that they are distinct isozymes and not different oligomeric species or aggregates of identical subunits.

Concluding Remarks

The biochemistry of *Spirulina* was previously reviewed by Ciferri (1983) and Ciferri and Toboni (1985). Although these reviews were published over ten years ago, the amount of information generated since then is fairly poor. Although a few unique physiological characteristics of *Spirulina* such as its alkalophilic nature were found, very little was done to study the biochemistry of the major metabolic activities. Little research has been carried out on the lipid and fatty acid metabolism of *Spirulina* (see Chapter 10). Some of the claims of the beneficial health properties of *Spirulina* are attributed to the relatively high content of γ-linolenic acid in the cells. What is the nature of this rather high content? Exploring the reason for the high rate of γ-linolenic acid accumulation may help not only in revealing the biochemistry of fatty acid metabolism in *Spirulina*, but may also have an impact on modifying the chemical composition as well as the selection of strains studied for production of specific chemicals.

There is no doubt that much more research has to be done. Development of genetic and molecular biology tools for *Spirulina* will greatly aid biochemical studies.

Acknowledgements

Much of the work presented in the section on growth and environmental stress was performed in the author's laboratory, in collaboration with his colleagues and graduate students. The following are to be thanked for their dedicated work: Dr G. Torzillo from Italy, Ms L. Chanawongse and Ms N. Kancharaksa from Thailand, Ms K. Hirabayashi from Japan, and Ms R. Guy and Ms N. Novoplansky from Israel.

References

BAKELS, R. H. A., VAN WALRAVEN, H. S., KRAB, K., SCHOLTS, M. J. C. and KRAAYENHOf, R. (1993) On the activation mechanism of the H^+-ATP synthase and unusual thermodynamic properties in the alkalophilic cyanobacterium *Spirulina platensis*, *Eur. J. Biochem.*, **213**, 957.

BLUMWALD, E. and TEL-OR, E. (1982) Osmoregulation and cell composition in salt adaptation of *Nostoc muscorum*. *Arch. Microbiol.*, **132**, 168.

BLUMWALD, E., MEHLHORN, R. J. and PACKER, L. (1984) Salt adaptation mechanisms in the cyanobacterium *Synechococcus* 6311. In SYBESMA, S. (Ed.) *Advances in Photosynthesis Research*, pp. 627–630, The Hague: Martinus Nijhoff/Dr W. Junk.

BOROWITZKA, L. J. (1986) Osmoregulation in blue-green algae. In ROUND, P. E. and CHAPMAN, O. J. (Eds), *Progress in Phycological Research*, Vol. 4, pp. 243–256, Biopress.

CARR, N. G. and WHITTON, B. A. (1973) *The Biology of Blue-Green Algae*. Botanical Monograph Volume 9, pp. 676, Oxford: Blackwell Scientific.

CIFERRI, O. (1983) *Spirulina*, the edible microorganism, *Microbiol. Rev.*, **47**, 551.

CIFERRI, O. and TIBONI, O. (1985) The biochemistry and industrial potential of *Spirulina*, *Ann. Rev. Microbiol.*, **39**, 503.

CORNET, J. F., DUSSAP, C. G. and DUBERTRET, G. (1992a) A structured model for simulation of cultures of the cyanobacterium *Spirulina platensis* in photobioreactors. I: Coupling between light transfer and growth kinetics. *Biotechnol. Bioeng.*, **40**, 817.

CORNET, J. F., DUSSAP, C. G., CLUZEL, P. and DUBERTRET, G. (1992b) A structured model for simulation of cultures of the cyanobacterium *Spirulina platensis* in photobioreactors. II: Identification of kinetic parameters under light and mineral limitations. *Biotechnol. Bioeng.*, **40**, 826.

CRITCHLEY, C. (1981) Studies on the mechanism of photoinhibition in higher plants, *Plant Physiol.*, **67**, 1161.

EHRENFELD, J. and COUSIN, J. L. (1984) Ionic regulation of the unicellular green alga *Dunaliella teritolecta*: Response to hypertonic shock. *J. Membr. Biol.*, **77**, 45.

FOGG, G. E. (1975) *Algal Cultures and Phytoplankton Ecology*, 2nd Edition, pp. 175, Wisconsin: The University of Wisconsin Press.

FORLANI, G., RICCARDI, G., DE ROSSI, E. and DE FELICE, M. (1991) Biochemical evidence for multiple forms of acetohydroxy acid synthase in *Spirulina platensis*, *Arch. Microbiol.*, **155**, 298.

FRY, I. V., HUFLEJT, M., ERBER, W. W. A., PESCHEK, G. A. and PACKER, L. (1986) The role of respiration during adaptation of the freshwater cyanobacterium *Synechococcus* 6311 to salinity, *Arch. Biochem. Biophys.*, **244**, 686.

GABBAY, R. and TEL-OR, E. (1985) Cyanobacterial biomass production in saline media. In PASTERNAK, D. and SAN PEITRO, A. (Eds) *Biosalinity in Action: Bioproduction with Saline Water*, pp. 107–116, Dordrecht: Martinus Nijhoff.

GREER, D. H., BERRY, J. A. and BJORKMAN, O. (1986) Photoinhibition of photosynthesis in intact bean leaves. Role of light and temperature, and requirement for chloroplast-protein synthesis during recovery, *Planta*, **168**, 253.

GUTERMAN, H., VONSHAK, A. and BEN-YAAKOV, S. (1989) Automatic on-line growth estimation method for outdoor algal biomass production, *Biotechnol. Bioeng.*, **34**, 143.

HAGEMANN, M., TECHEL, D. and RENSING, L. (1991) Comparison of salt- and heat-induced alterations of protein synthesis in the cyanobacterium *Synechocystis sp.* PCC 6803, *Arch. Microbiol.*, **155**, 587.

HICKS, D. B. and YOCUM, C. F. (1986) Properties of the cyanobacterial coupling factor ATPase from *Spirulina platensis*. II: Activity of the purified and membrane-bound enzymes, *Archives of Biochemistry and Biophysics*, **245**, 230.

IEHANA, M. (1983) Kinetic analysis of the growth of *Spirulina sp.* on continuous culture. *J. Ferment. Technol.*, **61**, 475.

IEHANA, M. (1987) Kinetic analysis of the growth of *Spirulina sp.* in batch culture. *J. Ferment. Technol.*, **65**, 267.

JENSEN, S. and KNUTSEN, G. (1993) Influence of light and temperature on photoinhibition of photosynthesis in *Spirulina platensis. J. Appl. Phycol.*, **5**, 495.

KAPLAN, A. (1981) Photoinhibition in *Spirulina platensis*: response of photosynthesis and HCO_3^- uptake capability to CO_2^- depleted conditions. *J. Exper. Bot.*, **32**, 669.

KOGUCHI, O. and TAMURA, G. (1988) Ferredoxin-sulfite reductase from a cyanobacterium *Spirulina platensis*, *Agric. Biol. Chem.*, **52**, 373.

63

KYLE, D. J. and OHAD, I. (1986) The mechanism of photoinhibition in higher plants and green algae. In STAEHELIN, L. A. and ARNTZEN, C. J. (Eds). *Encyclopedia of Plant Physiology*, New Series, Vol. 19, Photosynthesis III, pp. 468–475, Berlin: Springer-Verlag.

LEE, H. Y., ERICKSON, L. E. and YANG, S. S. (1987) Kinetics and bioenergetics of light-limited photoautotrophic growth of *Spirulina platensis. Biotechnol. Bioeng.*, **29**, 832.

LERMA, C. and GÓMEZ-LOJERO, C. (1987) Preparation of a highly active ATPase of the mesophilic cyanobacterium *Spirulina maxima, Photosynthesis Res.*, **11**, 265.

LEWIN, R. A. (1962) *Physiology and Biochemistry of Algae*, pp. 929, New York: Academic Press.

MARQUEZ, F. J., NISHIO, N., NAGAI, S. and SASAKI, K. (1995) Enhancement of biomass and pigment production during growth of *Spirulina platensis* in mixotrophic culture, *J. Chem. Tech. Biotechnol.*, **62**, 159.

MARQUEZ, F. J., SASAKI, K., KAKIZONO, T., NISHIO, N. and NAGAI, S. (1993) Growth characteristics of *Spirulina platensis* in mixotrophic and heterotrophic conditions. *J. Ferment. Bioeng.*, **76**, 408.

MARTEL, A., YU, S., GARCIA-REINA, G., LINDBLAD, P. and PEDERSÉN, M. (1992) Osmotic-adjustment in the cyanobacterium *Spirulina platensis*: Presence of a γ-glucosidase, *Plant Physiol. Biochem.*, **30**, 573.

MINAMI, Y., WADA, K. and MATSUBARA, H. (1989) The isolation and characterization of a cytochrome b_6f complex from the cyanobacterium *Spirulina sp., Plant Cell Physiol.*, **30**, 91.

MOLITOR, V., ERBER, W. and PESCHEK, G. A. (1986) Increased levels of cytochrome oxidase and sodiumproton antiporter in the plasma membrane of *Anacystis nidulans* after growth in sodium enriched media, *FEBS Lett.*, **204**, 251.

OGAWA, T. and AIBA, S. (1978) CO_2 assimilation and growth of a blue-green alga, *Spirulina platensis*, in continuous culture. *J. Appl. Chem. Biotechnol.*, **28**, 5151.

OGAWA, T. and TERUI, G. (1970) Studies on the growth of *Spirulina platensis* (I) On the pure culture of *Spirulina platensis. J. Ferment. Technol.*, **48**, 361.

ÖQUIST, G. (1987) Environmental stress and photosynthesis. In BIGGINS, J. (Ed.) *Progress in Photosynthesis Research*, Vol. IV, Dordecht: Martinus Nijhoff.

OWERS-NARHI, L., ROBINSON, S. J., DEROO, C. S. and YOCUM, C. F. (1979) Reconstitution of cyanobacterial photophosphorylation by a latent Ca^{+2}-ATPase, *Biochem. and Biophys. Res. Comm.*, **90**, 1025.

PIRT, S. J. (1975) *Principles of Microbe and Cell Cultivation*, Oxford: Blackwell.

POWLES, S. B. (1984) Photoinhibition of photosynthesis induced by visible light, *Annu. Rev. Plant Physiol.*, **35**, 15.

REED, R. H., RICHARDSON, D. L. and STEWART, W. D. P. (1985) Na^+ uptake and extrusion in the cyanobacterium *Synechocystis* PCC 6714 in response to hyper-saline treatment. Evidence for transient changes in plasmalemma Na^+ permeability, *Biochim. Biophys. Acta*, **814**, 347.

STEIN, J. R. (1973) *Handbook Phycological Methods. Culture Methods and Growth Measurements*, p. 448, Cambridge: Cambridge University Press.

TIBONI, O. and CIFERRI, O. (1983) Purification of the elongation factor Tu (EF-Tu) from the cyanobacterium *Spirulina platensis*, *Eur. J. Biochem.*, **136**, 241.

TORZILLO, G. and VONSHAK, A. (1994) Effect of light and temperature on the photosynthetic activity of the cyanobacterium *Spirulina platensis. Biomass and Bioenergy*, **6**, 399.

TORZILLO, G., SACCHI, A., MATERASSI, R. and RICHMOND, A. (1991) Effect of temperature on yield and night biomass loss in *Spirulina platensis* grown outdoors in tubular photobioreactors, *J. Appl. Phycol.*, **3**, 103.

VAN LIERE, L. and MUR, L. (1979) Growth kinetics of *Oscillatoria agardhii* Gomont in continuous culture limited in its growth by the light energy supply, *J. Gen. Microbiol.*, **115**, 153.

VONSHAK, A. (1986) Laboratory techniques for the culturing of microalgae. In Richmond, A. (Ed.) *Handbook for Algal Mass Culture*, pp. 117–145, Boca Raton, Fl.: CRC Press.

VONSHAK, A. (1991) Microalgae: Laboratory growth techniques and the biotechnology of biomass production. In HALL, D. O., SCURLOCK, J. M. O., BOLHÀR-NORDENKAMPF, H. R., LEEGOOD, R. C. and LONG, S. P. (Eds) *Photosynthesis and Production in a Changing Environment: A Field and Laboratory Manual*, pp. 337–355, London: Chapman and Hall.

VONSHAK, A. and RICHMOND, A. (1981) Photosynthetic and respiratory activity in *Anacystis nidulans* adapted to osmotic stress, *Plant Physiol.*, **68**, 504.

VONSHAK, A., GUY, R., POPLAWSKY, R. and OHAD, I. (1988a) Photoinhibition and its recovery in two different strains of *Spirulina*. *Plant and Cell Physiology*, **29**, 721.

VONSHAK, A., GUY, R. and GUY M. (1988b) The response of the filamentous cyanobacterium *Spirulina platensis* to salt stress. *Arch. Microbiol.*, **150**, 417.

XU, C., NAGIDAT, A., BELKIN, S. and BOUSSIBA, S. (1994) Isolation and characterization of plasma membrane by two-phase partitioning from the alkalophilic cyanobacterium *Spirulina platensis*, *Plant & Cell Physiol.*, **35**, 737.

YABUKI, Y., MORI, E. and TAMURA, G. (1985) Nitrite reductase in the cyanobacterium *Spirulina platensis*, *Agric. Biol. Chem.*, **49**, 3061.

ZARROUK, C. (1966) Contribution à l'étude d'une cyanophycée. Influence de divers facteurs physiques et chimiques sur la croissance et photosynthèse de *Spirulina maxima* Geitler. Ph.D. Thesis, University of Paris.

4

Genetics of *Spirulina*

AJAY K. VACHHANI AND AVIGAD VONSHAK

Introduction

Despite the general acceptance that classical genetics and modern genetic engineering techniques are powerful and almost essential tools in studying molecular regulation processes, these techniques have not been applied very efficiently to *Spirulina*. The need for new improved strains to be used for particular purposes is just another reason for the need to develop a gene transfer system and other molecular genetic techniques for *Spirulina*. When reviewing the available relevant literature on *Spirulina*, one is surprised to note how little has been done, even in comparison with the work done on other filamentous blue-green algae.

Indeed, genetic studies on cyanobacteria have been restricted to a few species, especially the unicellular ones which are easy to handle and mainly used for isolation of mutants or cloning of genes involved in photosynthesis. Another group includes the filamentous nitrogen fixers, which have been used in order to study gene regulation involved in the metabolic process of nitrogen fixation, or the developmental process by which vegetative cells are differentiated into heterocysts.

Some of the reasons for the scarcity of genetic studies using *Spirulina* as the experimental organism may be the following.

1 It is essential to use axenic (bacteria free) cultures. Only a few species have been isolated and maintained in this condition.

2 It is much easier to perform classical genetics and isolation of mutants with species that form colonies when plated on agar and have a good plating efficiency. Most of the *Spirulina* isolates plate at a relatively low efficiency and exhibit gliding motility on agar plates. This problem has been overcome recently in our lab and in some others, by reducing the salt concentration in the agar plates and by isolating strains that have lost their gliding ability.

3 It seems that most of the *Spirulina* species have a very high endonuclease and exonuclease activity, which makes the introduction of foreign DNA molecules very difficult.

Table 4.1 Amino acid analog resistant mutants of *Spirulina platensis*

Selected agents	Mutagen	No. of mutants isolated	Frequency	Reference
5-fluorotryptophan	NTG	1	1.1×10^{-7}	Riccardi et al., 1981a
β-2-thienylalanine	NTG	38	1.2×10^{-6}	
Ethionine	NTG	80	to	
p-fluorophenylalanine	NTG	27	7.1×10^{-6}	
Azetidine-2-carboxylic acid	NTG	145	per plated filament of 100 cells	
β-2-thienyl-DL-alanine	UV	–	$\geqslant 10^{-4}$	Lanfaloni et al., 1991
β-2-thienyl-DL-alanine	MNNG	–	$\geqslant 10^{-5}$	
8-azaguanine	MNNG	–	$\approx 10^{-5}$	
β-2-thienyl-DL-alanine	None (spontaneous)	–	3×10^{-7}	
8-azaguanine	None (spontaneous)	–	6×10^{-7}	

4 Endogenous plasmids, which are used in many microorganisms for construction of shuttle vectors, have not been detected in the *Spirulina* species checked.

So far, the major part of the work on *Spirulina* genetics has been focused on:

1 isolation and characterization of mutants (Table 4.1);
2 cloning and analysis of genes involved in fundamental processes such as protein synthesis, carbon dioxide fixation and nitrogen metabolism (Table 4.2);
3 attempts to construct a gene transfer system for *Spirulina*.

Mutagenesis and Isolation of Mutants

Almost all the work dealing with isolation of mutants was performed by Riccardi's group in Italy. Using various amino acid analogs, Riccardi et al. (1981a) reported the isolation of *Spirulina platensis* mutants resistant to 5-fluorotryptophan, β-2-thienyl-alanine, ethionine, p-fluorophenylalanine or azetidine-2-carboxylic acid. Nitroso-guanidine (NTG) was used as the mutagenic agent, and almost 300 resistant mutants were isolated. Some of the mutants appeared to be very frequent, e.g. more than 100 mutants resistant to azetidine-2-carboxylic acid and 80 resistant to ethionine were isolated.

A few of the mutants such as AZ8, PF27 and TA35 appeared to be resistant to more than one analog and to overproduce the corresponding amino acid. These mutants may carry mutations in the mechanisms regulating amino acid biosynthesis.

A second group consisted of mutants that were resistant to one analog only. From this group, one resistant to azetidine-2-carboxylic acid was found to overproduce only proline, while one resistant to fluorotryptophan and one resistant to ethionine did not overproduce any tested amino acid.

Table 4.2 Genes cloned from *Spirulina platensis*

Gene(s) for	Sequenced	Reference
1. Ribulose-1,5-bisphosphate carboxylase (large and small subunits)	No	Tiboni et al., 1984a
2. Glutamine synthetase	No	Riccardi et al., 1985
3. 32 kDa thylakoid membrane protein and α and β subunits of C-phycocyanin	No	De Rossi et al., 1985
4. Translation elongation factor (two genes)	No	Tiboni et al., 1984b
5. Str operon (ribosomal proteins S7 and S12 and translation elongation factors EF-G and EF-Tu)	Yes	Buttarelli et al., 1989
6. β-isopropylmalate dehydrogenase	Yes	Bini et al., 1992
7. Ribosomal protein S2 and part of the gene for elongation factor Ts (EF-Ts)	Yes	Sanangelantoni et al., 1990
8. Acetohydroxy acid synthase (genes for two isoenzymic forms)	Yes	Milano et al., 1992
9. Ribosomal protein S10	Yes	Sanangelantoni and Tiboni, 1993
10. Serine esterase	Yes	Salvi et al., 1994
11. 16S ribosomal RNA; 23S ribosomal RNA; transfer RNA-ile	Yes	Nelissen et. al., 1994
12. delta-12 desaturase (*des* A)	Yes	Deshnium, 1995
13. delta-6 desaturase (*des* D)	Yes	Murata et al., 1996
14. (3R)-hydroxymyristoyl acyl carrier protein dehydrase (*fabZ*) homolog gene	Yes (partial codons)	Los and Murata, 1995
15 ATPase gamma subunit	Yes	Steinemann and Lill, 1995
16 Recombination protein (*rec* A)	Yes	Vachhani and Vonshak, 1995
17. Allophycocyanin genes: *apcA*, *apcB*, *apcC*.	Yes	Anjard, 1996

The mutant overproducing proline is very interesting, because in many studies it has been suggested that proline may be acting as an osmoregulant in salinity-adapted blue-green algae (see Chapter 3). Indeed, when the proline-overproducing strains and control cultures were exposed to 0.3 to 0.9 M NaCl, a marked decrease in growth was observed in the control cultures while the overproducing cells maintained a significant part of the original growth capacity. Studies of this type can be cited to illustrate the potential of simple classical genetic approaches in studying different problems in growth physiology of *Spirulina*.

Riccardi et al. (1981b) further characterized two mutants resistant to ethionine (an analog of methionine) and reported that they had different mode of resistance. One such mutant, ET7, presumably carried a mutation affecting the mechanisms regulating amino acid biosynthesis. As compared with the wild type strain, ET7 overproduced methionine and other amino acids and did not take up significant quantities of methionine from the medium. In contrast, the mutant ET17 grew at the same rate as the parental strain and did not overproduce methionine but showed a reduction in the amount of amino acids incorporated into the protein.

It was reported by Riccardi et al. (1982) that the mutation responsible for ethionine resistance in strain ET17 most probably involves an altered methionyl-tRNA synthetase, since the methionyl-tRNA present in crude extracts of ET17 showed a reduced affinity for methionine and ethionine.

Spontaneous valine-resistant mutants of *S. platensis* were isolated by Riccardi et al. (1988), and preliminary characterization of three of them indicated that one (strain DR2) was defective in valine uptake and two (strain DR5 and DR9) carried alterations in a valine-mediated mechanism of synthesis of acetohydroxy acid synthase, the first common enzyme of the pathway.

Lanfaloni et al. (1991) standardized the conditions for isolation of 8-azaguanine or β-2-thienyl-DL-alanine resistant mutants of *Spirulina*. Optimal conditions were found to be 1–3 min UV irradiation and 30 min incubation with 50 µg MNNG/ml of trichomes derived from cultures entering stationary phase and sonicated for 10 s and 5 s, respectively. In this respect, from our experience and that of others, it is highly recommended to use fragmented filaments for the mutagenesis step. From other experiments it is also evident that it is important to use slow-growing cultures to obtain a high mutagenesis rate. This is most likely because fast-growing filaments contain more then one complete copy of the genome, thus reducing the chances of isolating the mutants produced, unless enough time is given for segregation to take place.

Cloning and Characterization of *Spirulina* Genes

Glutamine Synthetase

Riccardi et al. (1985) described the isolation of the gene for glutamine synthetase (*glnA*) from *Spirulina platensis*. Their approach was to use a *glnA* gene probe derived from *Anabaena* 7120 and hybridize it to *S. platensis* DNA which had been digested with various restriction endonucleases. The entire *glnA* gene of *S. platensis* was found to be located on an 8 kbp *Hind*III fragment. *Hind*III-cut *S. platensis* genomic DNA and plasmid pAT153 were ligated together and used to transform cells of the *Escherichia coli* strain ET8051, a mutant that carries a deletion of the *glnA* gene and flanking sequences and requires glutamine for growth.

Transformants were selected on a minimal medium and thus the *glnA* gene was isolated by functional complementation of an *E. coli* mutant.

Identity of the cloned *glnA* gene was confirmed by transcription and translation *in vitro* of the purified plasmid DNA. Analysis by SDS-polyacrylamide gel electrophoresis of the [S^{35}] labeled proteins, produced with an *E. coli* cell-free system, demonstrated that a radioactive band of ca. 51 000 daltons, which corresponded to the glutamine synthetase monomer, was evident only when plasmids bearing the appropriate cloned fragments from *Spirulina* or *Anabaena* were utilized.

Acetohydroxy Acid Synthase

Acetohydroxy acid synthase (AHS, EC 4.1.3.18) is the first common enzyme in the biosynthetic pathways leading to the synthesis of valine, isoleucine and leucine and has been demonstrated to be present in *S. platensis* in two isoenzymic forms (Forlani et al. 1991).

An *S. platensis* genomic library was shown to contain a 4.2 kbp *Cla* I fragment and a 3.2 kbp *Cla* I/*Sal* I fragment carrying the presumptive genes encoding the two isoenzymic forms of AHS. The 4.2 kbp (*ilvX*) and the 3.2 kbp (*ilvW*) fragments were subcloned in the plasmid vector pAT153 and it was determined that the *ilvX* gene was able to complement a suitable mutant of *E. coli*, while the *ilvW* gene supported poor growth of the same mutant (Riccardi et al. 1991). Milano et al. (1992) determined the complete nucleotide sequence of *ilvX* and *ilvW* and showed the presence of two reading frames of 1836 and 1737 nucleotides for *ilvX* and *ilvW*, respectively. The predicted amino acid sequences of the two isoenzymes, compared with the *Synechococcus* PCC7942 AHS enzyme and the large subunits of the *E. coli* AHS I, II and III isoenzymes, revealed a notable degree of similarity. Unlike AHS isoenzymes isolated from *E. coli* and *Salmonella typhimurium* which are tetramers consisting of two large and two small subunits, a small subunit has not been identified for either of the *S. platensis* AHS isoenzymes. Northern blot hybridization analysis demonstrated that the *ilvX* and the *ilvW* genes are transcribed to give mRNA species of approximately 2.15 kbp and 1.95 kbp, respectively.

Genes for the 32 kDa Thylakoid Membrane Protein and Phycocyanin

De Rossi et al. (1985) constructed a cosmid library of *Spirulina platensis* and identified genes involved in photosynthesis (large and small subunits of D-ribulose-1,5-bisphosphate carboxylase, 32 kDa thylakoid protein, α,β subunits of phycocyanin) and protein synthesis (elongation factors EF-Tu and EF-G).

Southern blotting and hybridization with the *Anabaena* 7120 32 kDa probe gave positive signals on different restriction fragments of two cosmids (pSpR6 and pSpE17), suggesting the presence, in *S. platensis*, of more than one gene for the 32 kDa protein, as has been reported for *Anabaena* 7120.

The entire cluster of genes coding for the α and β subunits of phycocyanin from *Agmenellum quadruplicatum* along with the gene for the 33 kDa phycobilisome protein was used as a probe, and one cosmid (pSpE31) was isolated by colony hybridization.

Ribulose-1,5-Bisphosphate Carboxylase

Tiboni et al. (1984a) reported the cloning of the genes for the large and small subunits of ribulose-1,5-bisphosphate carboxylase from *S. platensis*. The probe for the large subunit gene was an internal 1 kbp fragment from the *Chlamydomonas reinhardii* gene. The genes for the large and small subunits were found to be very closely located on a 4.6 kbp DNA fragment. Genes for both the subunits appeared to be expressed, albeit to a different extent, in minicells of *E. coli*. The amount of the large subunit produced in the bacterial host represented at least 10 per cent of the total protein.

β-Isopropylmalate Dehydrogenase

β-Isopropylmalate dehydrogenase (EC 1.1.1.85) is a key enzyme in the isopropylmalate (IMP) pathway, which is the most common route for leucine biosynthesis in many species. Bini et al. (1992) cloned the gene for β-isopropylmalate dehydrogenase from *S. platensis* by heterologous hybridization using the *Nostoc* UCD7801 *leuB* gene as a probe. The entire *leuB* coding region was sequenced along with 645 bp of the 5' flanking region and 956 bp of the 3' flanking region. An open reading frame of 1065 nucleotides, capable of encoding a polypeptide of 355 amino acids was identified. Comparison of the amino acid sequences published for corresponding proteins from either bacteria or yeast and the amino acid sequence deduced from the nucleotide sequence of the *S. platensis* gene revealed a homology of 45 per cent or more. Northern hybridization analysis revealed that the *S. platensis leuB* gene was transcribed as a single monocistronic RNA, approximately 1200 bases long.

Genes for Ribosomal Proteins and Elongation Factors

Tiboni et al. (1984b) used probes derived from the *tuf*A (elongation factor Tu) gene of *Escherichia coli* to detect homologous sequences on *Spirulina platensis* DNA. They reported the isolation of a 6 kbp fragment of *S. platensis* DNA which appears to contain two sequences homologous to the *E. coli* gene. Thus, *S. platensis* presumably contains two *tuf* genes.

The genes encoding ribosomal proteins S12 and S7, as well as the protein synthesis elongation factors Tu (EF-Tu) and G (EF-G) of *S. platensis*, were identified and cloned by Tiboni and Pasquale (1987). Gene expression for ribosomal protein S12 was determined by genetic complementation. *E. coli* HB101, a streptomycin resistant strain, when transformed with a plasmid bearing the putative S12 gene, showed a streptomycin sensitive phenotype demonstrating that the S12 protein is not only synthesized in the bacterial cell but also integrated in the bacterial ribosomes. Gene expression for the EF-Tu gene was determined by production of the protein in *E. coli* minicells. Buttarelli et al. (1989) reported the nucleotide sequence of a 5.3 kbp DNA fragment carrying the *str* operon (ca. 4.5 kbp) of *S. platensis*. The *str* operon includes the following genes: *rpsL* (ribosomal protein S12), *rpsG* (ribosomal protein S7), *fus* (translation elongation factor EF-G) and *tuf* (translation elongation factor Ef-Tu). Primary structures of the four gene products were derived from the nucleotide sequence of the operon and compared with the available corresponding structures from eubacteria, archaebacteria and chloroplasts. In almost all cases, extensive homology was found and the order was S12>EF-Tu>EF-G>S7. The largest homologies were usually found between the cyanobacterial proteins and the corresponding chloroplast gene products. No codon usage bias was detected in *S. platensis*.

The *E. coli* gene for ribosomal protein S2 was used by Sanangelantoni et al. (1990) as a probe to clone a 6.5 kbp region of the *S. platensis* genome. Sequence analysis showed that the fragment contained the gene for ribosomal protein S2 and a part of the gene for the elongation factor Ts (EF-TS). The arrangement of *rpsB* and *tsf* with a spacer region in between resembles the arrangement of these genes in *E. coli*.

Deduced amino acid sequence of the *rpsB* gene showed higher similarity with the *E. coli* (68.5 per cent) than with the tobacco chloroplast (39.3 per cent) S2 ribosomal protein. The authors point out that this finding is rather unexpected, since the endosymbiont hypothesis for the origin of plastids proposes that they arose from an ancestral photosynthetic prokaryote related to cyanobacteria. The deduced amino acid sequence of the second incomplete ORF which was located 51 bp downstream of the *rpsB* gene was found to be 50.6 per cent identical to the *E. coli* elongation factor Ts. No sequence similarity was observed in the spacer region present between the two genes in *S. platensis* and *E. coli*.

Sanangelantoni and Tiboni (1993) reported the cloning of the structural gene (*rps10*) encoding ribosomal protein S10 and also determined the location of the *rps10* gene relative to the *tuf* gene in *S. platensis*. Alignment of the predicted S10 sequence of *S. platensis* with the homologous sequences from cyanelles, bacteria, archaebacteria and eukaryotes revealed a high degree of sequence homology (74 per cent amino acid identity) with the cyanellar protein. The *rps10* gene of *S. platensis* is adjacent to the *str* operon genes, unlike the situation in *E. coli* where it is located in a different operon – the S10 operon.

Serine Esterase

The gene for serine esterase from *Spirulina platensis* was cloned, identified and expressed in *E. coli* (Salvi et al. 1994). The approach used was of shotgun cloning. Chromosomal DNA of *Spirulina* was isolated and digested with *Bgl*II. Three- to five-kbp fragments of genomic DNA were cloned into plasmid pPLc2833. *E. coli* HB101 cells carrying pcI875 were transformed and selection was made on the basis of utilization of tributyrin with and without induction of the lambda promoter. The primary structure of the esterase deduced from the DNA sequence displayed a 32 per cent identity with the sequence of carboxyl esterase of *Pseudomonas fluorescens*. The findings reported suggest that the esterase of *S. platensis* is a serine enzyme.

Attempts to Develop Gene Transfer Systems for *Spirulina*

The absence of plasmids in *Spirulina* (there have been only two preliminary reports so far) has been a major obstacle in the development of gene transfer systems. Any attempt made to transfer genes into *Spirulina* would require the following points to be tackled:

1 entry of DNA into *Spirulina*;
2 evasion of restriction digestion by the cyanobacterial nucleases;
3 stable maintenance of the DNA (either by an independent *ori* or by integration into the host genome);
4 expression of the selection marker.

Of the points listed above, the presence of restriction endonucleases may, perhaps, be the most formidable. These enzymes have been found in many cyanobacterial species, and Kawamura et al. (1986) reported the presence of three restriction endonucleases in *Spirulina platensis* subspecies *siamese*. Recently Tragut et al. (1995) have reported the identification of four restriction enzymes in the soluble protein fraction of *Spirulina platensis* strain pacifica. Specificities of these enzymes are listed in Table 4.3.

As mentioned earlier there have been two reports on the presence of plasmids in *Spirulina*. Qin et al. (1993) have reported isolation of CCC DNA of size 2.40 kbp and 1.78 kbp from *Spirulina* strains S_6 and F_3 respectively. Prof. Hiroyuki Kojima (Government Industrial Research Institute, Osaka, Japan) has reported (personal communication) the presence of megaplasmids in *Spirulina*. His group is presently working to confirm these results.

Experiments describing successful transformation of *Spirulina platensis* by electroporation have been reported by Cheevadhanarak et al. (1993) and Kawata et al. (1993).

Cheevadhanarak et al. (1993) used the plasmid pKK232-8 which bears a promoterless chloramphenicol acetyltransferase (CAT) gene. Shotgun cloning of *Spirulina* genomic DNA upstream of this gene resulted in expression of CAT activity due to the *Spirulina* promoter sequences, and such constructs were selected in *E. coli*. Plasmids with a functional CAT gene were electroporated into *Spirulina* and transformants selected on appropriate selective medium. Transformants were stably maintained under selection pressure for several generations. PCR analysis of transformant DNA with CAT based primer sequences revealed the presence of the CAT gene, and Southern hybridization experiments indicated that pKK232-8 derivatives had integrated into the *Spirulina* genome.

Kawata et al. (1993) reported transformation of *S. platensis*, with heterologous DNA by electroporation. Vector DNA of three forms was used: The *E. coli* plasmid vector pHSG397, its *Eco* RI linearized DNA and linearized pHSG379 flanked with random fragments of host DNA at both ends. Integrative transformation was evaluated by chromosomal integration of the chloramphenicol acetyltransferase (CAT) gene and was confirmed by PCR amplification using appropriate primers. A second marker used was *lacZ*, and the activity of β-galactosidase was assayed by MUG (4-methylumbelliferyl-β-galactoside) method with HPLC.

In both the reports discussed above, stability of the transformants has been a major problem and hence more work remains to be done in these systems. It is clear that any transformation system developed for *Spirulina* must be an efficient one that will allow not only a good degree of expression of the gene introduced but also be able to maintain itself in a fast-growing culture.

Table 4.3 Specificities of restriction enzymes of *Spirulina platensis* subspecies *siamese* and strain pacifica

Strain	Enzyme description		Specificity	Reference
S. platensis subspecies *siamese*	*Spl*I	New enzyme	C/GTACG	Kawamura et al., 1986
	*Spl*II	Isoschizomer of *Tth*111I	GACNNNGTC	
	*Spl*III	Isoschizomer of *Hae*III	GGCC	
S. platensis strain pacifica	*Spa*I	Isoschizomer of *Tth*111I	GACN/NNGTC	Tragut et al., 1995
	*Spa*II	Isoschizomer of *Pvu*I	CGAT/CG	
	*Spa*III	Isoschizomer of *Pvu*II	CAG/CTG	
	*Spa*IV	Isoschizomer of *Hind*III	AAGCTT	

Concluding Remarks

The fairly poor level of information on the genetics and molecular biology of *Spirulina* is eventually going to limit the rate at which new information on its cell-biology, physiology and biochemistry can be accumulated. Development of a reliable genetic engineering methodology will be a definite breakthrough in this field. It will open the way, not only for better basic research, but will enable a better selection and screening program for development of new and better strains to be used by the industry. Although it may still take some time before a reliable gene transfer system is developed, studies on *Spirulina* genetics must continue. Even without the existence of a gene transfer methodology, basic work can be done along two main lines.

1 By using classical methodology more mutants can be isolated. Beside the fact that they can be used for biochemical studies, they will be very useful in the future once the transformation system is available.

2 Isolation of genes involved in specific regulation processes or associated with specific requirements of the biotechnology industry may be performed. This can be done by:

  (a) using easily selected markers like genes encoding resistance to inhibitors or genes coding for production of easily detected products;

  (b) using sequence homology of already characterized genes from other prokaryotes;

  (c) using *Spirulina* DNA for complementation of mutants in *E. coli*, or single cell cyanobacteria that have a reliable transformation system.

Today, the molecular genetics of *Spirulina* seems a neglected field of research and any piece of information accumulated at this stage is undoubtedly of high value. It is our hope that this situation will improve, probably as a result of an increased level of funding by the biotechnology industry which is going to benefit from these developments.

References

ANJARD, C. (1996) Direct submission of DNA sequence, *GenBank accession # X95898*.

BINI, F., DE ROSSI, E., BARBIERATO, L. and RICCARDI, G. (1992) Molecular cloning and sequencing of the β-isopropylmalate dehydrogenase gene from the cyanobacterium *Spirulina platensis, J. Gen. Microbiol.*, **138**, 493.

BUTTARELLI, F. R., CALOGERO, R. A., TIBONI, O., GUALERZI, C. O. and PON, C. L. (1989) Characterization of the *str* operon genes from *Spirulina platensis* and their evolutionary relationship to those of other prokaryotes, *Mol. Gen. Genet.*, **217**, 97.

CHEEVADHANARAK, S., KANOKSLIP, S., CHAISAWADI, S., RACHDAWONG, S. and TANTICHAROEN, M. (1993) Transformation system for *Spirulina platensis*. In MASOJIDEK, J. and SETLIK, I. (Eds) *Book of Abstracts of the 6th International Conference on Applied Algology*, p. 109, Czech Republic, September.

DE ROSSI, E., RICCARDI, G., SANANGELANTONI, A. M. and CIFERRI, O. (1985) Construction of a cosmid library of *Spirulina platensis* as an approach to DNA physical mapping, *FEMS Microbiol. Letters*, **30**, 239.

DESHNIUM P. (1995) Direct submission of gene sequence, *GenBank accession # X86736*.

FORLANI, G., RICCARDI, G., DE ROSSI, E. and DE FELICE, M. (1991) Biochemical evidence for multiple forms of acetohydroxy acid synthase. In *Spirulina platensis*, *Arch. Microbiol.*, **155**, 298.

KAWAMURA, M., SAKAKIBARA, M., WATANABE, T., KITA, K., HIRAOKA, N., OBAYASHI, A., TAKAGI, M. and YANO, K. (1986) A new restriction endonuclease from *Spirulina platensis*, *Nucleic Acids Res.*, **14**, 1985.

KAWATA, Y., YANO, S. and KOJIMA, H. (1993) Transformation of *Spirulina platensis* by chromosomal integration. In MASOJIDEK, J. and SETLIK, I. (Eds) *Book of Abstracts of the 6th International Conference on Applied Algology*, p. 72, Czech Republic, September.

LANFALONI, L., TRINEI, M., RUSSO, M. and GUALERZI, C. O. (1991) Mutagenesis of the cyanobacterium *Spirulina platensis* by UV and nitrosoguanidine treatment, *FEMS Microbiol. Lett.*, **83**, 85.

LOS, D. A., and MURATA, N. (1995) Direct submission of DNA sequence, *GenBank accession # U41821*.

MILANO, A., DE ROSSI, E., ZANARIA, E., BARBIERATO, L., CIFERRI, O. and RICCARDI, G. (1992) Molecular characterization of the genes encoding acetohydroxy acid synthase in the cyanobacterium *Spirulina platensis*, *J. Gen. Microbiol.*, **138**, 1339.

MURATA, N., DESHNIUM, P. and TASAKA, Y. (1996) Biosynthesis of gamma-linolenic acid in the cyanobacterium *Spirulina platensis*. In HUANG, Y. and MILLES, D. E. (Eds) *Gamma-linolenic acid, metabolism and its role in nutrition and medicine*, p. 22, Champaign, Illinois: AOC Press.

NELISSEN, B., WILMOTTE, A., NEEFS, J. M. and DE WACHTER, R. (1994) Phylogenetic relationships among filamentous helical cyanobacteria investigated on the basis of 16S ribosomal RNA gene sequence analysis, *Syst. App. Microbiol.*, **17**, 206.

QIN, S., TONG., S., ZHANG, P. and TSENG, C. K. (1993) Isolation of plasmid from the blue-green alga *Spirulina platensis*, *Chin. J. Oceanol. Limnol.*, **11**, 285.

RICCARDI, G., SORA, S., and CIFERRI, O. (1981a) Production of amino acids by analog-resistant mutants of the cyanobacterium *Spirulina platensis*, *J. Bact.*, **147**, 1002.

RICCARDI, G., SANANGELANTONI, A. M., CARBONERA, D., SAVI, A. and CIFERRI, O. (1981b) Characterization of mutants of *Spirulina platensis* resistant to amino acid analogues, *FEMS Microbiol. Lett.*, **12**, 333.

RICCARDI, G., SANANGELANTONI, A. M., SARASINI, A. and CIFERRI, O. (1982) Altered methionyl-tRNA synthetase in a *Spirulina platensis* mutant resistant to ethionine, *J. Bact.*, **151**, 1053.

RICCARDI, G., DE ROSSI, E., VALLE G. D. and CIFERRI O. (1985) Cloning of the glutamine synthetase gene from *Spirulina platensis*, *Plant Mol. Biol.*, **4**, 133.

RICCARDI, G., DE ROSSI, E., MILANO, A. and DE FELICE, M. (1988) Mutants of *Spirulina platensis* resistant to valine inhibition, *FEMS Microbiol. Lett.*, **49**, 19.

RICCARDI, G., DE ROSSI, E., MILANO, A., FORLANI, G. and DE FELICE, M. (1991) Molecular cloning and expression of *Spirulina platensis* acetohydroxy acid synthase genes in *Escherichia coli*, *Arch. Microbiol.*, **155**, 360.

SALVI, S., TRINEI, M., LANFALONI, L. and PON, C. L. (1994) Cloning and characterization of the gene encoding an esterase from *Spirulina platensis*, *Mol. Gen. Genet.*, **243**, 124.

SANANGELANTONI, A. M. and TIBONI, O. (1993) The chromosomal location of genes for elongation factor Tu and ribosomal protein S10 in the cyanobacterium *Spirulina platensis* provides clues to the ancestral organization of *str* and S10 operons in prokaryotes, *J. Gen. Microbiol.*, **139**, 2579.

SANANGELANTONI, A. M., CALOGERO, R. C., BUTTARELLI, F. R., GUALERZI, C. O. and TIBONI, O. (1990) Organization and nucleotide sequence of the genes for ribosomal protein S2 and elongation factor Ts in *Spirulina platensis*, *FEMS Microbiol. Lett.*, **66**, 141.

STEINEMANN, D. and LILL, H. (1995) Sequence of the gamma-subunit of *Spirulina platensis*: a new principle of thiol modulation of F0F1 ATP synthase? *Biochim. Biophys. Acta.*, **86**, 1230.

TIBONI, O. and DI PASQUALE, G. (1987) Organization of genes for ribosomal proteins S7 and S12, elongation factors EF-Tu and EF-G in the cyanobacterium *Spirulina platensis*, *Biochim. Biophys. Acta*, **908**, 113.

TIBONI, O., DI PASQUALE, G. and CIFERRI, O. (1984a) Cloning and expression of the genes for ribulose-1,5-bisphosphate carboxylase from *Spirulina platensis*, *Biochim. Biophys. Acta*, **783**, 258.

TIBONI, O., DI PASQUALE, G. and CIFERRI, O. (1984b) Two *tuf* genes in the cyanobacterium *Spirulina platensis*, *J. Bact.*, **159**, 407.

TRAGUT, V., XIAO, J., BYLINA, E. J. and BORTHAKUR, D. (1995) Characterization of DNA restriction-modification systems in *Spirulina platensis* strain pacifica. *J. Appl. Phycol.*, **7**, 561.

VACHHANI, A. K. and VONSHAK, A. (1995) Direct submission of DNA sequence, *GenBank accession # U33924*.

5

Outdoor Mass Production of *Spirulina*: The Basic Concept

AVIGAD VONSHAK

The past ten years have witnessed a burst of activity relating to production of micro-algae for commercial purposes. From a modest beginning of *Chlorella* tablets in Japan in the late 1950s, new endeavors have emerged as specialized industries the world over aimed at producing health food, food additives, animal feed, biofertilizers and an assortment of natural products (Richmond 1986a, 1986b; Borowitzka and Borowitzka, 1988). The development of algal biotechnology is reviewed in the introduction to this volume. A more detailed account of the problems and day-to-day maintenance parameters involved in large-scale operation is given in Chapter 8. I have tried to describe briefly the history of *Spirulina* as a staple in human diet in the preface to this volume. The purpose of this chapter is to provide the reader with basic information on outdoor mass cultivation of algae, establishing a scientific ground to methods used in commercial production sites as well as suggesting improvements to obtain a higher output rate from the system, leading presumably to a reduction in production cost, thus making outdoor mass production of *Spirulina* more commercially feasible.

The Concept

The concept of algal biotechnology is basically the same as in conventional agriculture, namely the utilization of the photosynthetic machinery for the production of biomass to be used as a source of food, feed, chemicals and energy. The main advantages of culturing microalgae as a source of biomass are as follows.

1 Microalgae are considered to be a very efficient biological system for harvesting solar energy for the production of organic compounds via the photosynthetic process.

2 Microalgae are non-vascular plants, lacking (usually) complex reproductive organs, making the entire biomass available for harvest and use.

3 Many species of microalgae can be induced to produce particularly high concentrations of chosen, commercially valuable compounds, such as proteins, carbohydrates, lipids and pigments.

4 Microalgae are microorganisms that undergo a simple cell division cycle, in most cases without a sexual type stage, enabling them to complete their cell cycle within a few hours and making genetic selection and strain screening relatively quick and easy. This also allows much more rapid development and demonstration of production processes than with other agricultural crops.

5 For many regions suffering low productivity due to poor soils or the shortage of sweet water, the farming of microalgae that can be grown using sea or brackish water and marginal land may be almost the only way to increase productivity and secure a basic protein supply.

6 Microalgal biomass production systems can be easily adapted to various levels of operational or technological skills, from simple, labor-intensive production units to fully automated systems which require high investments.

The Process

The main components of the production process of microalgae biomass are presented in Figure 5.1, illustrating the major inputs and potential uses of the biomass produced. The process can be divided into two main steps:

1 growing the algal biomass, which involves the biological knowledge and the operational parameters;

2 the engineering aspects dealing with the reactor (pond) design, harvesting and processing of the biomass produced.

Successful algae production combines the biological insights of growing photoautotrophic microorganisms with the special requirements for reactor design appropriate for the process.

The biological understanding required includes the effects of interactions of environmental factors, such as light and temperature, as well as salinity, photoinhibition and dark respiration, on algal growth and productivity. The role of

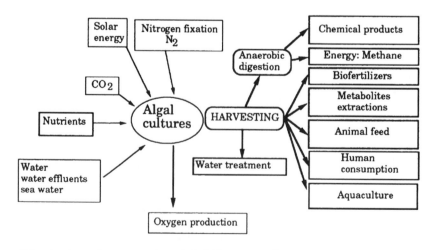

Figure 5.1 A schematic presentation of algal biomass production. Inputs and potential outputs.

these parameters in laboratory grown cultures is discussed in Chapter 3. They have to be considered in developing an operational protocol for pond management which also includes nutrient levels and pH, in order to establish a continuous culture for sustained production and to avoid development of grazers, predators and contamination by other algae.

Biological Problems and Limitations

The major biological limitations for production of *Spirulina* in mass cultures outdoors must be understood in order to obtain maximum output rate. These limitations may also serve as criteria for selecting outstanding strains of *Spirulina* to be used in outdoor production ponds. Assuming that the rate of photosynthesis can be used as an indication of the metabolic activity of outdoor algal cultures, we followed the daytime changes in oxygen concentration in the pond in order to correlate changes in oxygen concentration with diurnal changes in light and temperature. In what may be seen today as a somewhat naive interpretation, we initially concluded from the results presented in Figure 5.2 that in summer the main limiting factor for growth of *Spirulina* in outdoor cultures is light. This derives from the fact that the daily peak in oxygen concentration is reached at the same time that light intensity is maximum. In winter, however, the main limiting factor is temperature because of a shift in the peak of oxygen which follows the peak in the pond temperature rather than light intensity. The effect of temperature and light on growth of laboratory cultures was discussed in Chapter 3. As already stated, dealing with those two

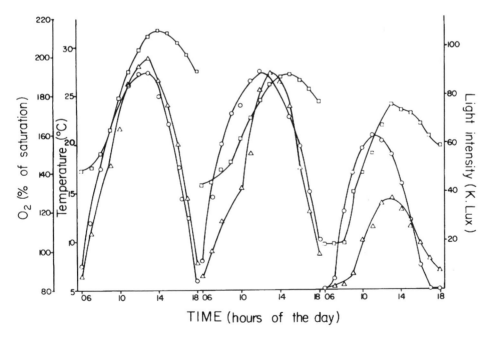

Figure 5.2 Diurnal changes in O_2 concentration, temperature and light during the course of representative days in different seasons. Left to right – summer, spring and winter. □—□ temperature in °C; o—o irradiance in klux; ▵—▵ O_2, percentage of saturation.

81

parameters separately may be somewhat simplistic. Nevertheless, since for many years they were considered as two separate limiting factors in outdoor mass cultivation of *Spirulina* (Vonshak, 1987a; Richmond, 1992b), only in recent studies has the degree of interaction and interrelationship between those factors and productivity been revealed (Vonshak, 1993).

The Effect of Light

Outdoor algal cultures are exposed to two rhythms of the light/dark regime. The first is relatively fast. It is induced by the mixing in the pond which results in a turbulent flow of the culture, dictating the frequency of the light/dark cycle (Laws et al., 1983). In this cycle algal cells are shifted between full solar radiation when located at the upper culture surface and complete darkness when reaching the bottom of the culture, usually at a depth of 12–15 cm. The time scale of such a cycle is measured in fractions of seconds. The other, relatively slower regime, is the change in solar irradiance during the day from sunrise to sunset. These two light cycles impose a unique physiological regime on the adaptation or acclimatization of outdoor algal cells to light.

Light limitation

When growing algae at a depth of 12–15 cm in open raceway ponds, self-shading governs the light availability to the single cell in the culture. Unless one uses a very diluted culture which allows penetration of light throughout the water column, a certain part of the culture will always fail to receive enough light to saturate photosynthesis. Thus almost by definition this kind of culture will be light limited. Indeed, in our very early studies (Richmond and Vonshak, 1978; Vonshak et al., 1982) we demonstrated that increasing cell concentration of the culture, which increases self-shading, results in a decrease of the growth rate. We carried out this kind of experiment during the summer, winter and spring, and findings indicated that the highest response of growth rate to cell concentration, i.e. self-shading, is observed in the summer (Figure 5.3). Our initial interpretation was that, in summer, temperatures are high enough, so the main limitation for growth of *Spirulina* outdoors is light. In winter and spring, however, when the temperature in the outdoor cultures is lower, the effect of self-shading is less pronounced (Figure 5.3).

As in many other microbial systems the important factor to be optimized is the output rate or productivity. The productivity of the system (Y) is defined as

$$Y = \mu X \tag{5.1}$$

where μ is the specific growth rate in units of reciprocal of time and X is the biomass concentration. As demonstrated by our earlier work (Richmond and Vonshak, 1978; Vonshak et al., 1982) and that of others (Richmond and Grobbelaar, 1986), there is an optimal biomass concentration which will correspond to the highest productivity. This concentration is not necessarily the one at which the highest μ is observed, again suggesting that conditions under which the highest output rate is obtained are light limited. This is further demonstrated in Figure 5.4 where the data of Figure 5.3 are replotted so that the output rate of the culture as a function of cell concentration is tested.

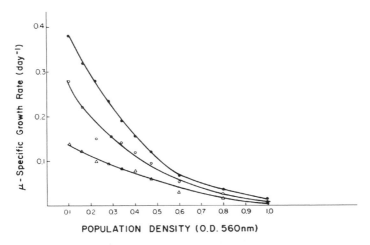

Figure 5.3 The specific growth rate of *Spirulina* as affected by population density and the seasons of the year: ●—● summer (June to September); o—o autumn and spring (October to November and April to May); ▵—▵ winter (December to February).

This observation points to one of the most important parameters in the management of outdoor cultures of *Spirulina*: the requirement to maintain an optimal cell concentration or what was defined later as an optimal areal density. In summer, when the main limiting factor is light, it can be seen (Figure 5.4) that any deviation from the optimal biomass concentration resulted in a significant reduction in the output rate of the system. It is also important to remember that the optimal concentration may also be affected by the culture depth, the strain used, and the rate of mixing. The last remains a topic of controversy.

Since the first reports (Richmond and Vonshak, 1978; Vonshak et al., 1982) on the effect of turbulent flow on the productivity of *Spirulina* ponds under outdoor conditions, two effects have been observed:

1 an increase in growth rate at the highest turbulence, resulting in an increase in productivity;

2 an increase in the optimal cell concentration at which the maximal productivity is obtained.

When this phenomenon was observed later by many others (Laws et al., 1983; Richmond and Grobbelaar, 1986; Grobbelaar, 1991), it was still a topic of debate (Grobbelaar, 1989). Interpretation ranged from claims that turbulent flow mimics the flashing light effect reported in the lab (Kok, 1953; Friedrickson and Tsuchiya, 1970), to claims that increased productivity is a result of a mass transfer phenomenon related to better uptake of nutrients and removal of toxic oxygen. We still believe that the effect of turbulent stirring has to do with the role of availability of light and its distribution in dense algal cultures. When stirring is insufficient, only a laminar flow is induced. As is the case for many large-scale commercial ponds, light distribution is unfavorable, leaving a significant part of the culture in complete darkness while another part is overexposed and may even suffer from photoinhibition (this phenomenon will be discussed in the next section). Thus, in a way, increasing the turbulent flow represents the most practical means of improving light distribution

Spirulina platensis (Arthrospira)

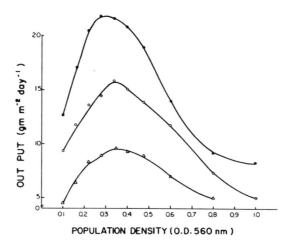

Figure 5.4 The effect of population density on the output rate through the seasons of the year: ●—● summer; o—o spring and autumn; ▵—▵ winter.

in outdoor cultures. Further support for our interpretation is suggested by the finding that, as a result of increased turbulence flow, not only is an increase in productivity observed, but also the highest production is achieved at a higher cell concentration (Figure 5.5).

Photoinhibition

Mainly under laboratory conditions, the phenomenon of photoinhibition has been well studied (Kyle and Ohad, 1986; Neale, 1987). For many years it was assumed that outdoor dense cultures, where light penetrates only part way, cannot be photoinhibited. Even when the phenomenon of photoinhibition was discussed as a factor in algal productivity, it related mainly to natural habitats as an ecological factor (Powles, 1984).

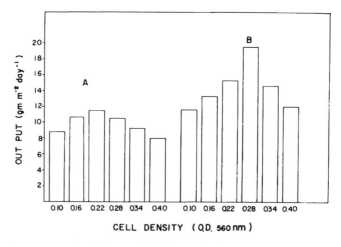

Figure 5.5 The effect of turbulent flow on the output rate of outdoor *Spirulina* cultures grown at different cell concentrations: (A), 7 rpm (slow); (B), 17 rpm (fast).

84

Vonshak and Guy (1988) were the first to describe the phenomenon of photoinhibition in outdoor-grown *Spirulina* cultures. By following the *in situ* photosynthetic activity of outdoor cultures grown at full solar radiation or under shaded conditions, they observed that shading the cultures resulted in an increase in photosynthetic activity and an increase in productivity.

These findings which seem somewhat to contradict the dogma that outdoor algal cultures are light limited, were then studied further. Using a more advanced methodology of variable chlorophyll fluorescence, it is possible to get a very fast and reliable indication of the quantum efficiency of photochemistry, to which the value of Fv/Fm is considered to be directly correlated. Furthermore, a reduction in the Fv/Fm ratio in many systems indicates photoinhibitory damage induced in PS II.

We have thus followed the Fv/Fm ratio in *Spirulina* cultures grown outdoors in shaded and non-shaded ponds. As demonstrated in Figure 5.6, a marked decline in

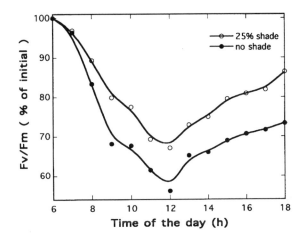

Figure 5.6 Daily changes in *Fv/Fm* in two *Spirulina* cultures grown outdoors, ●—● under full solar radiation, and o—o at 25 per cent cut-off by shading.

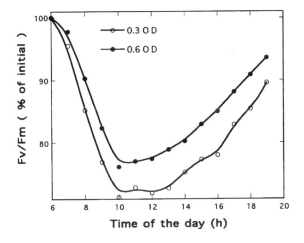

Figure 5.7 Daily changes in *Fv/Fm* in two *Spirulina* cultures grown outdoors at two cell concentrations: o—o 0.3 OD and ●—● 0.6 OD.

the ratio, reaching its lowest value at midday, has been observed in the cultures. When the pond was shaded so as to reduce light intensity by 25 per cent, the degree of inhibition was also significantly reduced. It is worth noting that once solar radiation lowers in the afternoon, a recovery in the PS II efficiency occurs. As already pointed out, light availability to the single cell in outdoor cultures is highly dependent on the cell concentration. We have compared the Fv/Fm ratio in outdoor cultures maintained at different cell concentrations. As seen in Figure 5.7, diluting the cell concentration of the culture and thus increasing the amount of light to which the cells are exposed, results in a higher degree of inhibition. This observation further supports our interpretation that the decline in Fv/Fm ratio is associated with an inhibition process caused by an excess exposure to light.

In the previous section, we suggested that increasing the turbulent flow in outdoor cultures represents a very practical way of improving the light regime. Using the parameter of Fv/Fm, we were able to demonstrate that this is indeed the case. When fast- and slow-mixed cultures are compared, the level of PS II efficiency in the fast-mixed culture is higher than that in the slow (Figure 5.8). We suggest that at least part of the reason for the higher productivity obtained at the higher turbulent/mixing system is a result of the prevention of a photoinhibitory stress that occurs in the slow-mixed cultures.

It seems that after almost 15 years of study related to the role of light in productivity of outdoor algal cultures, *Spirulina* in particular, we have reached a better understanding of the complicated light environment to which algal cells are exposed. We know that due to extreme shifts in the level of light intensity, at least in *Spirulina* cultures, photoinhibition may take place. The fact that photoinhibited *Spirulina* cultures have a lower photosynthetic efficiency (Figure 3.3) means that they require more light to reach the same level of activity as non-photoinhibited cells, thus making photoinhibited cultures actually light limited. This finally leads to what may be seen as the paradox of light in outdoor *Spirulina* culture: during a significant part of the day, the outdoor cultures are photoinhibited and light limited at the same time.

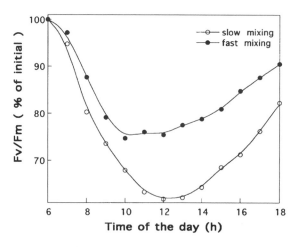

Figure 5.8 Daily changes in *Fv/Fm* of two *Spirulina* cultures grown outdoors at o—o slow and ●—● fast mixing rate.

The Effect of Temperature

In many regions of the world, temperature may represent the main limitation for high biomass production rates in outdoor open ponds of *Spirulina* cultures. Although the effect of temperature on growth rate of laboratory algal cultures is well documented, its effect in outdoor cultures is still not fully understood. An outdoor algal culture undergoes a diurnal cycle which in areas out of the tropics may show a difference of 20 °C. In the morning, the pond temperature may only be in the range 15–20 °C; an optimal temperature in the range 35–38 °C is reached only in the early afternoon. Even in the tropics where the culture does reach the optimal temperature, during a significant part of the day – the early morning hours – the temperature will still be much below the optimum.

One possible advantage to the diurnal cycle may be a low temperature at night. It has been demonstrated that relatively high temperatures at night can increase respiration rate, which may result in the phenomenon described as night loss of biomass. The degree of loss varies as a function of the biomass composition and may reach values of 30 per cent of the previous daily productivity (Torzillo et al., 1991; Guterman et al., 1989; Grobbelaar and Soeder, 1985).

During winter, *Spirulina* cannot be grown in outdoor open ponds, except in the tropics (see Chapter 8). The only way by which cultures can be maintained so as to overcome the low temperature, so that production can be resumed once winter is over, is to put them in greenhouses or to significantly increase the biomass concentration.

Temperature limitation on productivity may represent one of the many drawbacks of open raceway systems. (This is one of the big advantages of closed reactors.) Figure 5.9 illustrates and compares diurnal changes in temperature of three cultures of *Spirulina* in open ponds, a greenhouse, and a tubular reactor. As can be seen in closed systems, the optimal temperature is reached almost 4 h earlier than in open pond cultures. This not only increases the time during which photosynthesis may operate at maximum capacity but also prevents some inhibitory effects due to the interaction of low temperature and high light intensity, as discussed in the following section.

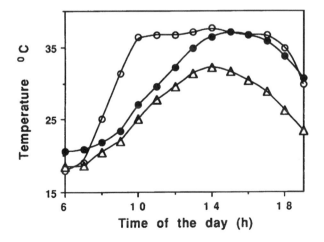

Figure 5.9 Diurnal temperature changes in three *Spirulina* cultures grown outdoors in △—△ open pond, ●—● open pond placed in a greenhouse, o—o a tubular reactor.

Spirulina platensis (Arthrospira)

Interaction of temperature and light

The possible interaction of temperature and light in outdoor algal cultures was overlooked for many years (Vonshak, 1987a; Richmond, 1992b), even though some of the early observations indicated that, when the pond temperature was increased by a few degrees in the morning, an increase in productivity much above what may be expected from just a simple temperature effect was observed. The special interaction of light and temperature stress in higher plants led us to examine the possibility that similar interaction took place in *Spirulina*. A recent paper (Vonshak et al., 1994) proposed that a relatively low morning temperature with rapid increase in light intensity may induce photoinhibitory stress. This was easily demonstrated in tubular reactors where fast heating is possible, and indeed, when the temperature of the outdoor *Spirulina* culture is maintained at 35 °C, the typical reduction in Fv/Fm does not occur (Figure 5.10).

In a much more detailed study, we have used *Spirulina* cultures grown in open ponds during the winter in Sede-Boker, Israel. In order to further elucidate the nature of the interaction between light and temperature, four experimental treatments were tested: two cultures were heated (up to 35 °C) during daylight, while in two other cultures the temperature, fluctuating between 5 °C in the early morning and a maximum of 20 °C midday, was not modified. One pond of each temperature treatment was shaded by a net, reducing by 25 per cent the total solar radiation. When following the changes in the Fv/Fm ratio in the four ponds during the day (Figure 5.11), a fast decline in Fv/Fm is observed, mainly in the non-heated ponds. The fastest decline, which takes place between 8 and 11 am, is slowed down to some extent when the culture is shaded. It has to be pointed out that light intensity at that time of the year is only 50–60 per cent of that measured during summer, reaching a maximum of 1400 $\mu mol\, m^{-2}\, s^{-1}$ at midday. Whereas heating the cultures significantly prevented the inhibitory effect observed in the non-heated cultures, shading the heated cultures did not provide any further protection. These results clearly indicate that early morning low temperatures in outdoor open cultures may cause photoinhibitory stress. Since the recovery from photoinhibitory stress is slower

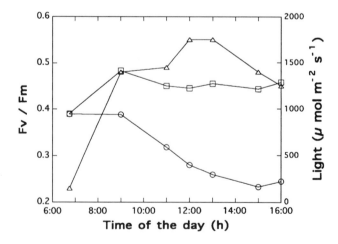

Figure 5.10 Changes in Fv/Fm in two *Spirulina* cultures grown outdoors at o—o 25 °C and □—□ 35 °C. The diurnal changes in light intensity (PAR) are indicated by ▵—▵.

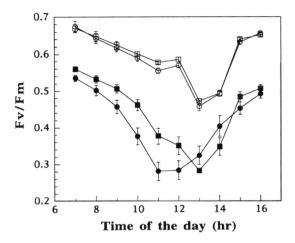

Figure 5.11 The effect of light and temperature on the *Fv/Fm* ratio of *Spirulina platensis* grown in open ponds under (□, ■) 25 per cent shading and (o, ●) non-shaded conditions during the daytime. □, o cultures were heated; ■, ● no heating applied.

than the rate at which the stress is induced, even if temperature is increasing later in the day, this early morning inhibition may result in lower productivity during most of the day. Following the daily productivity of the four ponds indicates a good correlation between the PS II efficiency and the daily output rate (Table 5.1). An increase in productivity in the two heated ponds occurs as expected. Shading the heated culture, which did not provide any protection to PS II, resulted in a small reduction in productivity by some 10 per cent. What is more striking is the effect of shading the non-heated cultures. These cultures responded to the 25 per cent reduction of light intensity by a marked increase in productivity of almost 45 per cent. It is thus clear that when *Spirulina* is exposed to suboptimal temperatures, the susceptibility of cells to photoinhibition is significantly increased.

It can be concluded that light and temperature are not separate factors that affect biomass productivity in a simple manner as in laboratory cultures. The fact that cultures in open ponds undergo diurnal and seasonal fluctuations makes this interaction somewhat more complicated to understand. The physiology of outdoor algal cultures is a new field of study, and much more has still to be done in order to fully reveal the interaction between environmental factors and production of outdoor *Spirulina* cultures.

Table 5.1 The effect of light and temperature on the productivity of outdoor *Spirulina* cultures

Experiment conditions	Production ($g\,m^{-2}\,day^{-1}$)
Heated	14.9
Heated + 25% shade	11.1
No heat, no shade	3.7
25% shade	5.3

Spirulina platensis (Arthrospira)

The Effect of High Oxygen Concentrations

Oxygen evolves in the process of photosynthesis and, as a result, accumulates in the culture. One important role of the turbulent flow, in addition to effecting a proper light-regime, is to remove that oxygen. In relatively small ponds, where high flow rates can be maintained, oxygen concentration can be at levels not higher than 200 per cent of air saturation, i.e. $12-14\ \mathrm{mg\,l^{-1}}$ depending on the temperature and barometric pressure. In large ponds, where the water flow is relatively slow (10 to $20\ \mathrm{cm\,s^{-1}}$), when high photosynthetic rates exist, the O_2 concentration may reach as high as 500 per cent of saturation. High concentrations of O_2 inhibit photosynthesis and growth and may lead to a total loss of the culture. Recently, the ill effects of high O_2 tension in *Spirulina* cultures were studied (Marquez et al., 1995; Singh et al., 1995). Both report that exposing laboratory cultures of *Spirulina* to high oxygen may result in reduced growth rate and bleaching of the pigments. Rather than artificially increasing the oxygen, we have tried to study the effect of high oxygen under outdoor conditions, by letting it accumulate as a result of photosynthesis. By measuring the fluorescence yield we could demonstrate that when the oxygen level is maintained at a range of 20 to $22\ \mathrm{mg\,l^{-1}}$ no significant reduction in the photosynthetic activity is observed. Exposing cultures to a higher level than this results in a rapid decline in the photosynthetic activity (Figure 5.12). This correlates with the daily output rate measured in the cultures, as well as with the observations that large-scale open pond levels of above 300 per cent of saturation also cause inhibitory effects on *Spirulina* productivity.

Maintenance of Monoalgal Cultures

Contamination by different algal species may present a very severe problem for microalgal cultures grown in outdoor open ponds. In a previous paper (Vonshak et al., 1983), we described a set of conditions that is instrumental in preventing

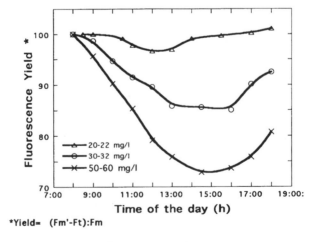

*Yield= (Fm'-Ft):Fm

Figure 5.12 Diurnal changes in the fluorescence yield of *Spirulina* cultures grown outdoors at 22 (▲), 32 (o) and 50–60 (×) $\mathrm{mg\,l^{-1}}$ of oxygen. *Fm'* and *Ft* are maximal and minimal fluroescence intensity of light-adapted cultures, respectively.

90

contamination of outdoor *Spirulina* culture by *Chlorella*. In most cases, the steps that proved effective in prevention of the *Chlorella* contamination were maintaining a high bicarbonate concentration (e.g. 0.2 M), taking precautions to keep the dissolved organic load in the culture medium as low as possible, and increasing winter temperature by greenhouse heating. Development of grazers in the culture, mainly the amoebae type, presents another problem. Vonshak et al. (1983) noticed amoebae grazing on *Chlorella*, and amoebae grazers of *Spirulina* were also observed in some improperly maintained commercial ponds. Addition of ammonia (2 mM) arrested the development of these grazers. Lincoln et al. (1983) also reported that the population of grazers was significantly reduced when ammonia was used as the main nitrogen source.

When a population of *Chlorella* cells growing in mass cultures of *Spirulina* increased in number (above $10^6 \, ml^{-1}$), repeating treatments with 1 mM ammonia was sufficient to prevent further proliferation of *Chlorella* (Vonshak and Boussiba,

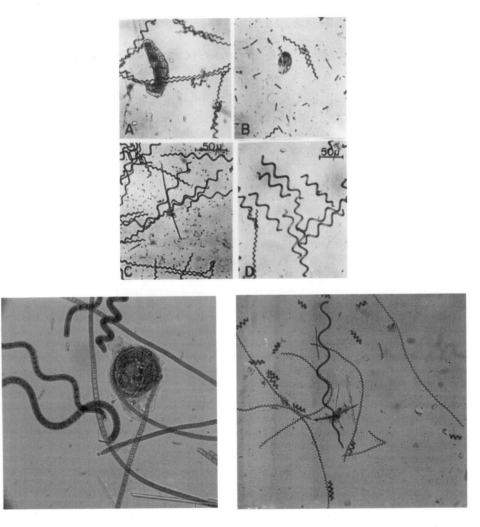

Figure 5.13 Photomicrographs showing different levels of contamination in outdoor cultures of *Spirulina* note the variation in *Spirulina* filaments as well).

unpublished results). Typical contamination and predators that may develop in *Spirulina* cultures are presented in Figure 5.13.

In conclusion, contamination of outdoor *Spirulina* cultures by other organisms can be controlled. Experience indicates that, as a rule, contaminating organisms do not present a serious difficulty as long as good growth is maintained in a monoalgal culture. It is worth noting that no cyanophages attacking *Spirulina* have been observed so far.

Engineering and Design Parameters

Some of the basic engineering and design parameters to be considered for the large-scale operation of algal ponds are summarized in the following section.

Design and Mixing of the Open-channel Raceway Ponds

Almost all commercial reactors for production of *Spirulina* are based on shallow raceways in which algal cultures sustained by a paddle wheel are mixed in a turbulent flow. One exception is Sosa Texcoco (Mexico) where a natural lake and certain facilities of a production plant for sodium bicarbonate were adapted for producing *Spirulina*. (This site has been recently closed down because of a strike, and it is not clear whether it will be re-open.) Recently another semi-natural lake in Myanmar (Burma) has been reported to be used as a production site for *Spirulina*. At other production sites in the USA, China, India, Thailand and Taiwan, two types of open raceway ponds are used: the first, which is more capital intensive, is lined with concrete (Thailand, India); the second is a shallow earthen tunnel lined with PVC or some other durable plastic. The cost and durability of the lining significantly influences the capital costs and thus the economic feasibility of this biotechnology. Benemann et al. (1987) estimated that any durable liner will add up to US $0.5 to the cost of production of each kg of algal biomass produced, demonstrating the need for cheaper lining such as low-cost clay sealing. Such lining has yet to be tested for durability under turbulent flow and periodic cleaning of the pond.

The size of commercial ponds varies from 0.1 to 0.5 ha. They are all stirred by paddle wheels, the design of which varies significantly from large wheels (diameter up to 2 m) with low revolution speed (10 rpm) to small wheels (diameter 0.7 m) with two to three times faster revolution. Culture depth is usually maintained at 15–18 cm. In most places it is determined by technical means, such as ground leveling and the degree of immersion of the paddle wheel.

Although the paddle wheel is the most common device for stirring *Spirulina* in commercial plants, other mixing devices are being tested. One difficulty with paddle stirring lies in the nature of the flow. It is usually insufficiently turbulent to effect an optimal light regime for the single cell (Vonshak et al., 1982; Richmond, 1987). Laws et al. (1983) introduced into the raceway an array of foils, a design similar to an airplane wing, which effected systematic mixing through the vortices created by the foils. These authors report a more than two-fold increase in photosynthetic efficiency. Others (Valderrama et al., 1987), following an idea originally suggested by Baloni et al. (1983), described a novel device for mixing shallow algal ponds. It consists of a board which closes the pond's cross-section, except for a slit above the bottom of the pond. The board is moved back and forth, creating a turbulent back whirl as the culture is forced through the slit. The authors claim to have achieved 'outstanding' results in growing *Spirulina*. This method of inducing turbulence in shallow raceways has not yet been scaled up, and a comprehensive evaluation of the system has yet to be carried out. Cultivating *Spirulina* in tubular reactors whose flow is induced by a suitable pump seems another promising means of cultivation which has yet to be tested on a large scale. Initial work by Tomaselli et al. (1987) argues for the potential of such a system (see Chapters 6 and 7).

Harvesting and Processing of Spirulina *Biomass*

In all commercial production processes, similar filtration devices are used for harvesting. These are basically of two types of screen: inclining or vibrating. Inclined screens are 380–500 mesh with a filtration area of 2–4 m² per unit and are capable of harvesting 10–18 m³ of *Spirulina* culture per hour. Biomass removal efficiency is high, up to 95 per cent, and two consecutive units are used for harvesting up to 20 m²h⁻¹ from which slurry (8–10 per cent of dry weight) is produced. Vibrating screens can be arranged in double or triple decks of screens up to 72 inches (183 cm) in diameter. Vibrating screens filter the same volume per unit time as the inclining screens, but require only one-third the area. Their harvesting efficiencies are often very high. At one commercial site a combination of an inclining filter and a vibrating screen has been used. Two main problems exist with the systems described above. In the process of pumping the algal culture to be filtered, the filaments of *Spirulina* may become physically damaged, and repeated harvesting leads to an increasingly enriched culture with unicellular microalgae or short filaments of *Spirulina* that pass through the screen readily.

The slurry (8–10 per cent of dry weight) obtained after filtration is further concentrated by filtration using vacuum tables or vacuum belts, depending on the production capacity. This step is also used for washing excess salts from the biomass; it amounts to 20–30 per cent of dry weight. The washed cake is frequently homogenized before being dried. This is accomplished by spray- or drum-drying. The end product should have an ash content of ca. 7 per cent. For good preservation and storage, moisture should not exceed 3–4 per cent.

Production of *Spirulina* by Simple Technology and Locally Available Inputs

Production of *Spirulina* may be greatly simplified, thereby avoiding the high-technology systems described so far. Such a mode of production results in relatively low output rates, which are compensated for by the much-reduced cost of production. Indeed, *Spirulina* culture lends itself readily to simple technology: cultivation may be carried out in unlined ditches through which flow is low (e.g. 10 cm s⁻¹). Stirring may be provided by a simple device driven by wind energy or harnessed to humans. Harvesting may be readily performed using some suitable cloth, and the biomass dehydrates in the sun. The quality of the *Spirulina* product obtained in this fashion would not be as high as what is attained in 'clean cultures', but the product could serve very well as animal feed. In the work of Granoth and Porath (1984), *Tilapia mossambica* was cultivated in artificial ponds with relatively high stocking density and fed with a mixture of solar-dried *Spirulina*, that had been cultivated and processed using low-cost technology and added to groundnut cake. The resulting mean food conversion ratio was lower than that observed using control fish fed with the usual fish ration. Furthermore, the yield of *Tilapia* fed on *Spirulina* mixed with groundnut cake was 41 per cent higher than that of fish fed on groundnut cake alone. (For more on the use of *Spirulina* in aquaculture see Chapter 11.)

Becker and Venkataraman (1982) and Seshadri and Thomas (1979) suggested a *Spirulina* growth medium based on low-cost nutrients obtained from rural wastes such as bone meal, urine or the effluent from biogas digesters. Indeed, this last possibility, an integrated system making use of effluent from a starch production factory using tapioca, was developed and scaled up to a demonstration plant by the team of King Mongkut's Institute of Technology in Bangkok. In this plan, a 160 m^3 digestor is operated for production of biogas with four 200 m^2 algal ponds that produce *Spirulina* biomass which is tested for nutritional value. The system has been scaled up to 14 ponds of 1000 m^2 each and an annual production of 30–50 tons. In another work carried out by Ms Jiamjit Boonsom from the National Inland Fisheries Institute, *Spirulina* strains, mainly from the north-eastern part of Thailand, were isolated. Those strains capable of growing in brackish water were mixed with casava or fish meal and served as a locally produced protein source for fish feed. Elsewhere, Bai (1986) has summarized the know-how developed for production of *Spirulina* biomass in a village in India. According to Chung et al. (1978), *Spirulina* grows well in diluted fermented swine manure, provided that the concentration of ammonia nitrogen is adjusted to 100 $mg\,l^{-1}$ and proper nutrients are supplemented. More detailed information on the use of wastewater for growing *Spirulina* biomass is given in Chapter 9.

Spirulina may also be grown on sea water enriched with urea, after excess Ca^{2+} and Mg^{2+} are precipitated (Faucher et al., 1979). In a recent report (Tredici et al., 1986) the use of sea water for cultivating *Spirulina* without any pre-treatment of the water is described. The bicarbonate concentration was about one-tenth lower than the usual, the pH was maintained at 9.0, and the medium was not recycled. Yields of 10 $g\,m^{-2}\,day^{-1}$ were obtained, pointing to the possible use of sea water as a cheap medium for the mass cultivation of *Spirulina*.

Economics and Future Prospective

Cost and Economic Evaluation

Many attempts have been made to predict the production cost of algal biomass produced on a large scale (Benemann et al., 1987; Dynatech, 1978). Rather than present another economic calculation, we mean to elaborate the economic basis of commercial production plants and indicate reasons for the large discrepancy between the actual cost of production and that which was predicted by some early calculations. The main difficulty stems from lack of information caused by commercial secrecy. We estimate that in the intensive production sites (Thailand, USA) the cost of production (excluding capital costs) is in the range of US$7–10 per kg of spray-dried *Spirulina* powder (some of those estimates are summarized in Table 5.2). Manpower represents 20–30 per cent of the running cost, a major

Table 5.2 Major items of investment and production costs in a *Spirulina* production site. Total pond area 50 000 m² (intended for high-value food grade production)

Item	Cost (thousand $US)
Investment costs	
Land preparation and development; site operation	138
Water and power network	257
Buildings (labs, offices, shops)	79
Nutrients, storage and stock	50
Pond: including lining, pump, mixing	493
Harvesting: including filtering, drying, packing	541
TOTAL	1558
Annual production costs	
Manpower	320
Repair and maintenance	42
Fixed operation costs	163
Variable operating costs (gas, nutrients, power, etc.)	123
Administration, capital and depreciation	390
TOTAL	1038

component. This reflects one of the main difficulties in attempting to cut down costs. Even in the biggest production sites in the USA, consisting of ten 0.5-ha ponds, the total production area is still too small to realize significant reduction of manpower cost per unit product. According to recent publications this site has doubled the number of ponds in one recent year (1995). According to some published economic analyses, minimal plant size must be no less than 10 to 100 ha, consisting of 1- to 10-ha ponds. Such large production sites are yet to be constructed and tested, so the matter of the effect of size of production on the cost of unit product remains open. Another open question is the effect of increased supply and reduced selling price on consumer demand.

A second major component in the cost of production is the cost of nutrients, particularly carbon. It ranges from 15 to 25 per cent of the total operation costs. The main reason for this high cost is the relatively low efficiency of the conversion of nutrients to algal mass. When nitrogen is supplied in the form of liquid ammonia, a significant part may be lost to the atmosphere in the form of ammonia gas, NH_3. This is due to the high pH (ca. 10.0) used for the cultivation of *Spirulina*. This problem is overcome when nitrate, NO_3^-, is used as the nitrogen source, but it is significantly more expensive than ammonia. Loss of nitrate through denitrification may take place especially in large ponds with low mixing rates, in whose still corners anaerobic pockets may be found. In one commercial plant, we calculated a nitrate loss of up to 50 per cent. The carbon requirement is usually supplied with CO_2, following an initial supply of 0.2 M sodium bicarbonate (Vonshak, 1986). The highest efficiency of CO_2 conversion into biomass obtained in commercial reactors of *Spirulina* is about 80 per cent. This high figure can be obtained in relatively small operational units (0.1–0.2 ha) in which exact monitoring and control systems are in place. In most of the production sites where one carbonation system is used for large ponds (0.3–0.5 ha), lower efficiencies are obtained. When growing *Spirulina* at a pH level higher than 9.6, some of the carbon requirements are met by atmospheric CO_2 (BenYaakov et al., 1985).

Current low efficiency of nutrient utilization also stems from a lack of knowledge concerning the use of recycled medium for long periods of time. Depending upon the production site and local experience, the medium must be changed completely three to six times a year in order to sustain production and to avoid deterioration of the culture. In at least one case, reuse of this 'low-quality' medium for so-called 'feed grade biomass' is practised. Clearly, subsequent information concerning recycling of the culture medium may result in less frequent changes of expensive nutrient medium, cutting costs significantly. In most cases, energy represents some 15 per cent of the operational cost, reflecting the requirement of power for mixing, pumping and drying. A breakthrough in the economic utilization of solar and/or wind energy would have to take place for this cost to decline significantly. In order to make *Spirulina* biomass a widely used commodity, it is clear that the cost of production must be reduced to the range of US$4–6 per kg of dry matter. This figure will be attained only if much higher production rates, about twice as high as at present, are commercially achieved. Recent market information indicates that while the cost of *Spirulina* production has fallen in some sites to about US$8 per kg, the price of powder sold in the market is more than US$20 per kg. One of the reasons for this high price may stem from the increase in demand when Sosa Texcoco, which was supplying about 200–300 tonnes of *Spirulina*, ceased production in 1994 because of management and strike problems.

The economic analysis of low-technology endeavors based on local resources is far more complicated as no exact figures for estimating the cost benefit of locally produced proteins are available. The cost of manpower may become less relevant when one considers social factors such as added jobs, improvement in the standard of living, and prevention of migration of small farmers to urban centers.

Future Prospects

The future of algal biotechnology rests, to a large extent, on two factors: (a) the ability to reduce costs of production and thus make algal biomass a commodity traded in large quantities, rather than limited to the health food market; (b) the development of suitable reactors. Closed systems offer several advantages over open raceways. In closed systems, cultures are better protected from contaminants, and thus the maintenance of monoalgal cultures should be easier. Water loss and the ensuing increase in salinization of the medium are much reduced. This mode of production opens the possibility of using sea water with low bicarbonate concentrations, thus saving on the cost of water and medium. Because of much higher cell densities, areal volumes may be much smaller, thereby reducing harvesting costs. Finally, optimal temperatures may be established and maintained more readily in closed systems, resulting in higher output rates. The latter is an essential aspect in the production of a microorganism such as *Spirulina* whose growth temperature is 37 °C. All of these developments have yet to be tested on a large scale in order to evaluate whether the higher investment cost is indeed compensated for by higher annual yield.

In the last five years many research groups in France, Italy, Israel, Singapore and Australia have reported progress in their attempts to scale up experimental tubular reactors for mass production of different species of algae. In our lab, a 1000 l tubular reactor for the cultivation of *Spirulina* has been operating. The information gathered so far indicates that *Spirulina* can be grown at a standing biomass of 3–5 gl^{-1} with a daily productivity of 1 gl^{-1}. For more details on closed systems, see Chapters 6 and 7.

Summary

In the past decade, much progress has been made in developing appropriate biotechnology for microalgal mass cultivation aimed at establishing a new agro-industry. This chapter presents a number of basic requirements needed to achieve high productivity and low-cost production with this new agrotechnology. The first is the availability of a wide variety of algal species and strains that will favorably respond to a variety of outdoor conditions. Another essential requirement is an appropriate bioreactor, either by improving an existing open raceway type (Soeder, 1980), or by developing a new type, such as tubular closed systems (Gudin and Chaumont, 1984; Lee, 1986). The latter solution seems more promising. These developments must overcome the main limitation confronted by the industry, i.e. overall low sustainable productivity. The areal yields obtained today fall too short of the theoretical maximum, and attempts must be made to improve efficiency. But even more important than obtaining high efficiency of solar energy conversion is to design a system which will operate at a constant, sustainable rate of production.

Acknowledgements

Much of the information in the section related to outdoor growth and mass cultivation derives from collaborative work performed by the author in the microalgal biotechnology lab in the Jacob Blaustein Institute for Desert Research, Ben-Gurion University of the Negev, Sede-Boker, Israel together with his colleagues Amos Richmond and Sammy Boussiba. Others who collaborated in part of the work deserve thanks for their dedicated work: Dr G. Torzillo from Italy and Mr Supat Laorawat from Thailand. They have all made the last twenty years of outdoors work with *Spirulina* richly rewarding and a great pleasure.

References

BAI, J. N. (1986) Mud pot cultures of the alga *Spirulina fusiformis* for rural households, *Monograph ser. on Engineering of Photosynthesis Systems*, Shri AMM Murugappa Chettiar Res. Cent. Madras, India, **19**, 1–39.

BALONI, W. G., FLORENZANO, A., MATERASSI, R., TREDICI, M., SOEDER, C. J. and WAGNER, K. (1983) Mass culture of algae for energy farming in coastal deserts. In STURB, A., CHARTIER, P. and SCHELESER, G. (Eds) *Energy from Biomass*, pp. 291–295, London: Applied Science.

BECKER, E. W. and VENKATARAMAN, L. V. (1982) *Biotechnology and Exploitation of Algae: The Indian Approach*, Eschborn: German Agency for Technical Cooperation.

BENEMANN, J. R., TILLETT, D. M. and WEISSMAN, J. C. (1987) Microalgae biotechnology, *Trends in Biotechnol.*, **5**, 47.

BEN-YAAKOV, S., GUTERMAN, H., VONSHAK, A. and RICHMOND, A. (1985) An automatic method for on-line estimation of the photosynthetic rate in open algal ponds, *Biotechnol. Bioeng.*, **27**, 1136.

BOROWITZKA, M. and BOROWITZKA, L. (1988) *Microalgal Biotechnology*, Cambridge: Cambridge University Press.

CHUNG, P., POND, W. G., KINGSBURG, J. M., WALKER, E. F., JR. and KROOK, L. (1978) Production and nutritive value of *Arthrospira platensis*, a spiral blue-green alga grown on swine wastes, *J. Animal Sci.*, **47**, 319.

Dynatech R/b Comp., Cambridge, Massachusetts (1978) Cost analysis of aquatic biomass systems, prepared for US Dept. of Energy, Washington DC, HCP/ET-4000 78/1.

FAUCHER, O., COUPAL, B. and LEDUY, A. (1979) Utilization of seawater-urea as a culture medium for *Spirulina maxima*, *Can. J. Microbiol.*, **25**, 752.

FRIEDRICKSON, A. G. and TSUCHIYA, H. M. (1970) Utilization of the effects of intermittent illumination on photosynthetic microorganisms. In *Prediction and Measurement of Photosynthetic Productivity*, pp. 519–541, Wageningen Centre for Agricultural Publishing and Documentation.

GRANOTH, G. and PORATH, D. (1984) An attempt to optimize feed utilization by *Tilapia* in a flow-through aquaculture. In FISHELZON, E. (Ed.) *Proceedings of the International Symposium on Tilapia and Aquaculture*, Nazareth, Israel, pp. 550–558.

GROBBELAAR, J. U. (1989) Do light/dark cycles of medium frequency enhance phytoplankton productivity? *J. Appl. Phycol.*, **1**, 333.

GROBBELAAR, J. U. (1991) The influence of light/dark cycles in mixed algal cultures on their productivity, *Bioresource Technology*, **38**, 189.

GROBBELAAR, J. U. and SOEDER, C. J. (1985) Respiration losses in green alga cultivated in raceway ponds, *J. Plankton Res.*, **7**, 497.

GUDIN, C. and CHAUMONT, D. (1984) Solar biotechnology and development of tubular solar receptors for controlled production of photosynthetic cellular biomass for methane production and specific exocellular biomass. In PAIZ, W. and PIRRWITZ, D. (Eds) *Energy From Biomass*, pp. 184–193, Dordrecht: Reidel.

GUTERMAN, H., VONSHAK, A. and BEN-YAAKOV, S. (1989) Automatic on-line growth estimation method for outdoor algal biomass production, *Biotechnol. Bioeng.*, **132**, 143.

KOK, B. (1953) Experiments on photosynthesis by *Chlorella* in flashing light. In BURLEW, J. S. (Ed.) *Algal Culture from Laboratory to Pilot Plant*, pp. 63–158, Washington DC: Carnegie Institution of Washington.

KYLE, D. J. and OHAD, I. (1986) The mechanisms of photoinhibition in higher plants and green algae. In STAEHELIN, L. A. and ARNTZEN, C. J. (Eds) *Encyclopedia of Plant Physiology*, Vol. 19, pp. 468–475, Berlin: Springer-Verlag.

LAWS, E. A., TERRY, K. L., WICKMAN, J. and CHALUP, M. S. (1983) A simple algal production system designed to utilize the flashing light effect, *Biotechnol. Bioeng.*, **25**, 2319.

LEE, Y. K. (1986) Enclosed bioreactor for the mass cultivation of photosynthetic microorganisms: the future trend, *Trends in Biotechnol.*, **4**, 186.

LINCOLN, E. P., HALL, T. W. and KOOPMAN, B. (1983) Zooplankton control in mass algal cultures, *Aquaculture*, **32**, 331.

MARQUEZ, F. J., SASAKI, K., NISHIO, N. and NAGAI, S. (1995) Inhibitory effect of oxygen accumulation of the growth of *Spirulina platensis*, *Biotechnol. Lett.*, **17**, 225.

NEALE, P. J. (1987) Algal photoinhibtion and photosynthesis in aquatic environment, in KYLE, D. J., OSMOND, C. B. and ARNTZEN, C. J. (Eds) *Photoinhibition: Topics in Photosynthesis*, Vol. 9, pp. 39–65, Amsterdam: Elsevier.

POWLES, S. B. (1984) Photoinhibition of photosynthesis induced by visible light, *Annu. Rev. Plant Physiol.*, **35**, 15.

RICHMOND, A. (1986a) *Handbook of Microalgal Mass Culture*, Boca Raton, Fl.: CRC Press.

RICHMOND, A. (1986b) *Microalgal Culture*, CRC Critical Reviews in Biotechnology, **4**, 369.

RICHMOND, A. (1987) The challenge confronting industrial microalgal culture; high photosynthetic efficiency in large-scale reactors, *Hydrobiologia*, **151**, 117.

RICHMOND, A. (1992a) Mass culture of cyanobacteria. In MANN, N. H. and CARR, N. G. (Eds) *Photosynthetic Prokaryotes*, pp. 181–210, New York: Plenum Press.

RICHMOND, A. (1992b) Open systems for the mass production of photoautotrophic microalgae outdoors: physiological principles, *J. Appl. Phycol.*, **4**, 281.

RICHMOND, A. and GROBBELAAR, J. U. (1986) Factors affecting the output rate of *Spirulina platensis* with reference to mass cultivation. *Biomass*, **10**, 253–264.

RICHMOND, A. and VONSHAK, A. (1978) *Spirulina* culture in Israel, *Arch. für Hydrobiol.*, **11**, 274.

SESHADRI, C. V. and THOMAS, S. (1979) Mass culture of *Spirulina* using low-cost nutrients, *Biotechnol. Lett.*, **1**, 287.

SINGH, D. P., SINGH, N. and VERMA, K. (1995) Photooxidative damage to the cyanobacterium *Spirulina platensis* mediated by singlet oxygen, *Current Microbiol.*, **31**, 44.

SOEDER, C. J. (1980) Massive cultivation of microalgae: results and prospects, *Hydrobiologia*, **72**, 197.

TOMASELLI, L., TORZILLO, G., GIOVANNETI, L., PUSHPARAJ, B., BOCCI, F., TREDICI, M., PAPUZZO, T., BALLONI, W. and MATERASSI, R. (1987) Recent research on *Spirulina* in Italy, *Hydrobiologia*, **151**, 79.

TORZILLO, G., SACCHI, A. and MATERASSI, R. (1991) Temperature as an important factor affecting productivity and night biomass loss in *Spirulina platensis* grown outdoors in tubular photobioreactors, *Bioresource Technol.*, **38**, 95.

TREDICI, M. R., PAPUZZO, T. and TOMASELLI, L. (1986), Outdoor mass culture of *Spirulina maxima* in sea-water, *Appl. Microbiol.*, **24**, 47.

VALDERRAMA, A., CARDENAS, A. and MARKOVITS, A. (1987) On the economics of *Spirulina* production in Chile with details on dragboard mixing in shallow ponds, *Hydrobiologia*, **151**, 71.

VONSHAK, A. (1986), Laboratory techniques for the culturing of microalgae. In RICHMOND, A. (Ed.) *Handbook for Microalgal Mass Culture*, pp. 117–145, Boca Raton, Fl.: CRC Press.

VONSHAK, A. (1987a) Biological limitations in developing the biotechnology for algal mass cultivation, *Science de L'Eau.*, **6**, 99.

VONSHAK, A. (1987b) Strain selection of *Spirulina* suitable for mass production, *Hydrobiologia*, **151**, 75.

VONSHAK, A. (1993) Outdoor algal cultures: are they light limited, light saturated or light inhibited (a personal confession or confusion). In MASOJIDEK, J. and SETLIK, I. (Eds) *Progress in Biotechnology of Photoautotrophic Microorganisms*, Book of Abstracts, 6th International Conference on Applied Algology, Czech Republic, p.56.

VONSHAK, A. and GUY, R. (1988) Photoinhibition as a limiting factor in outdoor cultivation of *Spirulina platensis*. In STADLER, T., MOLLION, J., VERDUS, M.-C., KARAMANOS, Y., MORVAN, H. and CHRISTIAEN, D. (Eds) *Algal Biotechnology*, pp. 365–370, London: Elsevier Applied Science.

VONSHAK, A., ABELIOVICH, A., BOUSSIBA, S., ARAD, S. and RICHMOND, A. (1982) On the production of *Spirulina* biomass: effects of environmental factors and of the population density, *Biomass*, **2**, 175.

VONSHAK, A., BOUSSIBA, S., ABELIOVICH, A. and RICHMOND, A. (1983) Production of *Spirulina* biomass: Maintenance of monoalgal culture, *Biotechnol. Bioeng.*, **25**, 341–351.

VONSHAK, A., TORZILLO, G. and TOMASELI, L. (1994) Use of chlorophyll fluorescence to estimate the effect of photoinhibition in outdoor cultures of *Spirulina platensis*, *J. Appl. Phycol.*, **6**, 31.

6

Tubular Bioreactors

GIUSEPPE TORZILLO

List of Symbols

b	bend factor in resistance to water recycle $(-)$
D	internal diameter of tubing (m)
f	water flow rate in the tubular reactor $(m^3 s^{-1})$
g	gravitational acceleration (ms^{-1})
h	water head in reactor (m)
L	length of the reactor (m)
K'	consistency index (dynes $s^n cm^{-2}$)
N_{Re}	Reynolds number $(-)$
N'_{Re}	generalized Reynolds number $(-)$
N	number of bends in the loop reactor $(-)$
n'	flow behavior index
Pc	power required for water recycle (W)
s	shear factor in resistance in water recycle $(-)$
Tc	liquid recycle time between aeration stations (h)
V	liquid velocity in the tubular reactor (ms^{-1})
X	biomass concentration $(g l^{-1})$
Yo	oxygen yield from biomass produced $(-)$
μ	specific growth rate (h^{-1})
ρ	density of liquid $(kg\ m^{-3})$
η	viscosity of the liquid $(kg\ m^{-1} s^{-1})$

Introduction and Historical Background

Tubular reactors are most likely to be used as the next generation of enclosed algal culture system. They allow: (1) effective illumination due to a better surface-to-volume ratio, (2) the attainment of high biomass concentration, (3) effective sterilization of the reactor, (4) high efficiency of CO_2 conversion because of very low CO_2 losses by outgassing, (5) low levels of contamination, (6) easy monitoring and control of operational parameters so as to achieve better utilization of solar energy at all times. Tubular photobioreactors are constructed with either collapsible or rigid tubes.

The various enclosed photobioreactors were reviewed by Lee (1986). The first studies on the outdoor culture of *Chlorella* with plastic and glass tubings were carried out in the early 1950s by Davis and co-workers (1953) at the Carnegie Institution of Washington. The choice of a closed system was based on the conviction that it would allow easy cleaning and prevent contamination. The reactor had a culture capacity of 1 liter, 65 per cent of which was contained in the tubing section, with the remaining part in the settling chamber. The culture was recycled by a finger pump at a flow rate of 540–620 ml min^{-1}. The daily biomass increase of *Chlorella* culture grown outdoors was within the 0.5–1.3 g l^{-1} range of dry weight. Davis and co-workers concluded that both glass and plastic tubes are suitable for outdoor culture of *Chlorella*. Another outdoor tubular reactor was constructed at about the same time in Japan (Tamiya et al., 1953). A *Chlorella ellipsoidea* culture was circulated in a series of glass tubes 3 cm in diameter, connected with rubber tubing and immersed in a water basin in order to control the temperature. The culture was circulated by a motor-driven pump at a linear velocity of 15 to 30 cm s^{-1}. The culture was aerated with CO_2-enriched air in an indoor gas-exchange tower. A pilot-plant project was begun in 1951 at Arthur D. Little, Inc., Cambridge, Massachusetts, to study the feasibility of the production of *Chlorella* in large, thin-walled polyethylene tubes (Anonymous, 1953). Stability of plastic tubing, circulation device and cooling of culture were investigated. The results obtained with that pilot-plant allowed a practical estimate of the production costs of algae (Fisher, 1956). The conclusion was that the cost of dry product obtained in closed systems was higher than that of conventional equivalent materials. Cooling of the culture was found to be the most important cost factor. Complete elimination of this item would have reduced the estimated investment by almost 50 per cent, so that the cost of algal production would have been about the same as with open systems. Moreover, during the experiments carried out by Tamiya (1956) at about the same period, both in closed and open systems, no essential difference was found between the two culture systems as regards to the possibility of contamination of the culture by other organisms. These findings led to an almost general abandonment of closed bioreactors for mass culture of algae, while open systems gained importance. The design of the tubular reactor remained static for about 30 years until the early 1980s when Pirt and co-workers (1983) developed a theory for the design and performance of a tubular bioreactor for culturing *Chlorella*. The reactor was arranged in the form of a loop made from 52 m of glass tubing of 1 cm bore, covering about 0.5 m^2. The innovation introduced was the use of a film of photosynthetic culture, 1 cm in depth, in contrast to the 15 cm or more generally used in open ponds. The culture was recycled either by a peristaltic pump or an air-lift system. The effective surface-to-volume ratio was 127 m^{-1}, thus a biomass concentration of 20 g l^{-1} could be maintained in a continuous flow culture, allowing a reduction of operating costs and risks of contamination.

A closed photobioreactor for mass culture of *Porphyridium* was developed in 1983 by Gudin and Chaumont at Centre d'Etudes Nucléaires de Cadarache, France (Gudin and Chaumont, 1983). The reactor was made of polyethylene tubes of 64 mm diameter and 1500 m length. A double layer of polyethylene tubes was used. The culture was contained in the upper layer of tubes. The temperature control of the culture was achieved by either floating or submerging the upper layer on or in a pool filled with water by adjusting the amount of air in the tubes of the lower layer. The pilot plant facility was composed of five identical units of 20 m^2 each; the total culture volume was 6.5 m^3.

Florenzano and co-workers (Torzillo et al., 1986), at Centro di Studio dei Microrganismi Autotrofi of Florence, Italy, pioneered the development of a tubular bioreactor for the production of cyanobacteria, particularly *Spirulina*. It consisted of flexible polyethylene sleeves 14 cm in diameter and 0.3 mm thick. Later on, owing to their low mechanical strength, these were replaced by Plexiglas® tubes 13 cm in diameter and 4 mm thick (Figure 6.1). The reactor was made up of several tubes laid side by side on a white polyethylene sheet and joined by PVC (polyvinylchloride) bends to form a loop, each bend incorporating a narrow tube for oxygen release. At the exit of the tubular loop the culture suspension flowed into a receiving tank. A diaphragm pump raised the culture to a feeding tank containing a siphon that allowed discharges of about 340 l into the photobioreactor at 4 min intervals, sustaining a flow velocity of 0.26 ms^{-1}. The maximal length of the loop was 500 m, corresponding to a volume of 8 m^3 and to a surface area of 80 m^2. On average, under the climatic conditions of central Italy, the annual yield of *Spirulina* biomass obtained with such a tubular bioreactor was equivalent to 33 t dry wt ha^{-1} year^{-1}, while the yield of the same organism cultivated in open ponds was about 18 t ha^{-1} year^{-1}. This considerable potential improvement was due to a better culture temperature profile realized in the tubular bioreactor throughout the diurnal cycle and the year round. A controlled culture system for studying the growth physiology of photosynthetic microorganisms outdoors in tubular systems was set up in Florence in 1981 (Bocci et al., 1987). It was composed of eight thermostated photobioreactors and four air-conditioned metallic

Figure 6.1 General view of tubular bioreactors made of Plexiglas® tubes 7.4 cm and 13.1 cm i.d. operated by Centro di Studio dei Microrganismi Autotrofi, CNR, Florence, Italy.

shelters housing the control equipment and analysis instruments. The bioreactor (length 23 m, i.d. 4.85 cm) was placed in a stainless steel basin containing thermostated demineralized water. The culture was recycled by a PVC pump. The distance between blades and casing was 1.3 cm to minimize mechanical damage to the *Spirulina* filaments. The working volume of each reactor was 51 l. In 1991, another tubular bioreactor design for outdoor culture of *Spirulina* was devised in the Florence laboratory (Torzillo et al., 1993). It consisted of a 245 m long loop made of Plexiglas® tubes (i.d. = 2.6 cm, o.d. = 3.0 cm) arranged in two planes (Figure 6.2). Each tube in the lower plane was placed in the vacant space between two tubes in the upper plane. That way the two planes received all the incident radiation falling on the face of the reactor. The culture flowed from one plane to the other so that the two planes formed a single loop. Another important aspect of this photobioreactor design was the use of two air-lifts for the culture recycle, one for each plane of tubes. They facilitated both mixing of the culture and oxygen degassing. The reactor contained 145 l of culture. The effective illuminated surface-to-volume ratio was 49 m^{-1}. The mean daily net productivity of a *Spirulina platensis* culture achieved in July (mean solar radiation of 25.5 MJ m^{-2} day^{-1}) was 28 g m^{-2} day^{-1}. This biomass yield corresponded to a net volumetric productivity of 1.5 g l^{-1} day^{-1}.

Recently Richmond and co-workers (1993) have reported a tubular bioreactor for outdoor culture of *Spirulina* in Israel. It consists of (a) an air-lift pump, (b) a gas separator and (c) transparent reactor tubing running in parallel and connected by manifolds (Figure 6.3). The reactor pipes are thin-walled extruded tubing of transparent polycarbonate (3.2 cm o.d., 3.0 cm i.d.). Such a design has the advantage

Figure 6.2 General view of a two-plane air-lift tubular bioreactor made of Plexiglas® tubes 2.6 cm i.d. devised by Centro di Studio dei Microrganismi Autotrofi, CNR, Florence, Italy.

Figure 6.3 A tubular bioreactor installed at Microalgal Biotechnology Laboratory, Sede-Boker, Israel. (From Richmond et al., 1993, with permission.)

of reducing the amount of minor head losses compared with bioreactors arranged in the form of a loop in which the tubes are connected by U-bends, making it easier to scale-up in an industrial scale unit. Tubular photobioreactors are usually placed horizontally, but in recent years pilot vertical 'biocoil' facilities have been installed in the UK (Luton) and Australia (Borowitzka and Borowizka, 1989). The solar receptor is arranged as a coil of approximately 30 mm diameter low-density polyethylene tubing wound around an open circular framework (Figure 6.4). Temperature control is achieved by a heat exchange unit installed between the reactor and the pump. A 1300 l unit would have an approximate photosynthetic surface area of 100 m^2. With the biocoil it is possible to operate a *Chlorella* culture at a biomass concentration of up to 10 g l^{-1} without yield decline (Robinson, 1993). Finally, inclined tubular reactors have also been investigated for outdoor cultivation of microalgae (Lee and Low, 1991) and cyanobacteria (Hoshino et al., 1991) and are reviewed separately in this volume (Chapter 7).

Tubular Bioreactors: Design Criteria

The basic function of a properly designed bioreactor is to provide a controlled environment in order to achieve optimal growth and/or product synthesis with the particular microorganism employed. As a result, to achieve a successful bioreactor design, a background both in chemical engineering and biological sciences, such as microbiology, biochemistry and molecular biology, is required. The first important

Figure 6.4 Biocoil of polyvinylchloride tubes (3 cm i.d.) set on a mobile base working in Southern Ireland. (Photograph by kind permission of Dr L. F. Robinson.)

criterion is a basic understanding of the microbiology of the particular microorganism, i.e. its morphology and nutritional requirements for optimal growth. The geometrical configuration of the bioreactor and the characteristics of the circulating device are also important. With shear-sensitive organisms, the circulating device design should produce gentle mixing and a very low shearing effect. The viscosity of the culture and whether it behaves as a Newtonian or non-Newtonian fluid, will have a profound influence on mixing as well as on mass and heat transfer. The bioreactor design and its hydrodynamic characteristics must be tailored differently to suit the microorganism to be cultivated. The experience gathered over many years of studies performed in the author's laboratory, as well as in others', on the outdoor culture of *Spirulina* in tubular reactors, indicates the main factors that affect their performance. They are: (1) diameter of the tubing, (2) length of tube reactor, (3) mixing of the culture, and (4) circulating device.

Optimal Tube Diameter of the Bioreactor

The choice of the tube diameter represents an important issue for an optimal design of the bioreactor since it affects: (1) its light uptake; (2) the biomass concentration at which the culture can be operated and, consequently, the harvesting cost; (3) the daily volumetric productivity; (4) the concentration of oxygen in the culture; (5) the CO_2 storage capacity of the bioreactor; (6) the temperature course of the culture; (7) the flow pattern and head loss for culture recycling in the bioreactor. The basic requirement for the photoautotrophic growth of microalgae outdoors is the efficient harvesting of solar radiation. A bioreactor should thus have a large surface-to-volume ratio. This can be attained in bioreactors made of small-diameter tubes. In such bioreactors, the optimal temperature for growth is reached early in the morning by direct gain of solar heat and thereafter maintained at the optimum by spraying water, or using some other cooling device. In contrast, in the evening, the relatively low volume cools down quickly to the ambient night temperature, resulting in lower respiratory losses (Torzillo et al., 1991). The productivity of *Spirulina* grown outdoors in photobioreactors of different tube diameters is summarized in Table 6.1. As expected in a light-limited system, the smaller the diameter, the higher the volumetric productivity. Although, in principle, the photobioreactor should be built with tubes of small diameter, the use of very narrow tubes, in the concept of large-scale equipment involving long runs of tubing, may result in hydrodynamic problems with filamentous microorganisms such as *Spirulina*. Indeed, it can produce a relatively high viscosity suspension, even at a low biomass concentration. In addition, the quick O_2 build-up in the bioreactor can impose severe limitations to its length as discussed below. Some bioreactor designs may work extremely well in the laboratory but become impossible or very inefficient to operate when scaled-up. The experience gathered till now indicates that a tube diameter of 2.6–3.0 cm could be optimal for an industrial bioreactor of *Spirulina* (Torzillo et al., 1993; Richmond et al., 1993; Borowitzka and Borowitzka, 1989).

Optimal Length of Tubes

Since the CO_2 storage capacity of *Spirulina* medium is high, due to the high alkalinity of the growth medium and high pH optimum of the organism, the main factor that determines the tube length is oxygen build-up. Once the cells are pumped into the tubes of the bioreactor, the photosynthetic activity of the cells causes an increase in the oxygen concentration, at a rate that depends on light intensity, biomass concentration and temperature. Dissolved oxygen concentrations equivalent to four to five times air saturation can be easily reached in summer days. There is some evidence that an oxygen concentration over $30 \, mg \, l^{-1}$ has a negative influence on both the growth and the protein synthesis in *Spirulina* (Torzillo et al., 1984; Tredici et al., 1992). Consequently, the time cycle of the reactor (i.e. the run duration between two aeration stations), which depends on the length of the reactor tubing, must be chosen in such a way as to prevent the oxygen concentration in the

Table 6.1 Productivity of *Spirulina* obtained in photobioreactors having different tubing diameters (data from Centro di Studio dei Microrganismi Autotrofi, Florence, Italy)

i.d. of tubing (mm)	S/V ratio (m^{-1})	Optimal biomass concentration $(g \, l^{-1})$	Volumetric productivity $(g \, l^{-1} \, day^{-1})$
131	9.7	0.6	0.23
74	17	1.2	0.40
26	49	3.5	1.50

bioreactor from exceeding 30 mg l^{-1}. The oxygen production of an algal culture can be expressed according to the following equation (Pirt et al., 1983):

$$O_2 \text{ production } (gl^{-1}) = \frac{\mu XTc}{Yo} \qquad (6.1)$$

where μ is the growth rate of the culture (h^{-1}), X is the biomass concentration (gl^{-1}), Tc is the time cycle of the reactor (h), Yo (dimensionless) is the ratio between the biomass synthesized and the oxygen produced. For *Spirulina* grown with nitrate as the nitrogen source, $Yo = 0.507$ (i.e. 1.97 g of oxygen are produced for 1 g of biomass). Table 6.2 shows the permissible combinations of culture speed and tube length for a photobioreactor made of tubes of i.d. 2.6 cm operated with a *Spirulina* culture. However, this data is based on the assumption that the oxygen inlet is near to the air saturation which, in general, would be an underestimation when air is used instead of nitrogen for oxygen degassing. The oxygen inhibition may be counteracted by increasing the speed of the culture. Alternatively, the length of the tubing should be reduced to prevent O_2 concentration from reaching the inhibitory level. The experience with *Spirulina* has confirmed that the optimal time cycle of a loop reactor of 2.6 cm inner diameter is about 6 min, which corresponds to a length of about 100 m at a culture speed of 0.3 m s^{-1} (Torzillo et al., 1993). Richmond et al. (1993) came to a similar conclusion with a bioreactor made with tubes of i.d. 3.0 cm for outdoor culture of *Spirulina*.

Mixing and Flow Parameters

Mixing of the culture is necessary (1) to ensure that all the cells are exposed to the light, (2) to maintain nutrient supply throughout the tubular reactor, (3) to diminish the nutritional and gaseous gradients surrounding the cells in actively growing cultures, improving the rate of exclusion of cell excretions, including O_2, at the cell surface. Stirring represents the most practical means by which efficient utilization of solar energy can be achieved (Richmond, 1990). When stirring is insufficient, the solar energy utilization declines, since the pattern of flow becomes laminar. Knowledge of the rheological behaviour of the culture is therefore very important for the design, scale-up and operation of the bioreactor, as it affects the culture mixing and mass transfer conditions in it. It has been demonstrated that *Spirulina* cultures behave as a Newtonian fluid at a biomass concentration of about 2 gl^{-1}. Over 4 gl^{-1} the cultures display decreasing viscosity with increasing shear rate, showing a pseudoplastic behavior (Figure 6.5). As a result, a lower power is required for pumping the culture in order to obtain a given increase in the speed. The non-

Table 6.2 Permissible combinations of culture speed and length of 2.6 cm i.d. tubing to maintain oxygen concentration below the inhibitory level of 30 mg l^{-1} for *Spirulina*

V (m/s)	0.2	0.3	0.6	0.8
L (m)	80	120	240	320

Notes: $\mu = 0.03$ h^{-1}; $X = 3.5$ gl^{-1}; $T = 35\,°C$

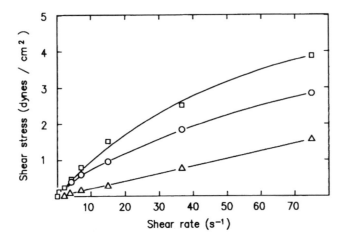

Figure 6.5 Shear stress versus shear rate diagram for *Spirulina* cultures having different biomass concentrations: (Δ) 2.3 gl^{-1}; (O) 5.0 gl^{-1}; (□) 6.74 gl^{-1}. (From Torzillo, 1993, unpublished results.)

Newtonian behavior could be a consequence of the filamentous morphology of this microorganism (Skelland, 1967). Indeed, dense cultures of *Spirulina* are probably characterized by extensive entanglement when the culture is at rest. Progressive disentanglement should occur under the influence of shearing forces, the filaments tending to orient themselves in the direction of the shear. According to Metzner and Reed (1955), the degree of turbulence for a non-Newtonian fluid is expressed with the generalized Reynolds number:

$$N'_{Re} = \frac{D^{n'} V^{(2-n')} \rho}{K' 8^{(n'-1)}}$$
(6.2)

where D is the tube diameter (m), V is the culture speed (m s^{-1}), ρ is the culture density (kg m^{-3}), n' is the flow behavior index (dimensionless) which defines the degree of non-Newtonian behavior of the culture, and K' (consistency index, dynes s^{n} cm^{-2}) defines its consistency; i.e. the larger the value of K' the 'thicker' or 'more viscous' the fluid. For Newtonian fluids, $n' = 1.000$ and K' reduces to η (the viscosity of the fluid), and N'_{Re} (Equation 6.2) reduces to the familiar $DV\rho/\eta$, showing that this traditional dimensionless group is merely a special restricted form of the more general one proposed by Metzner and Reed (1955). As *Spirulina* cultures behave as a non-Newtonian fluid at high biomass concentrations (over 4.0 gl^{-1}), i.e. like pseudoplastic fluids (n' less than unity), the onset of turbulence is expected to occur according to the results of Dodge and Metzner (1959), at a Reynolds number slightly greater than that of 2100 for Newtonian fluids. Furthermore, their results show some evidence that the Reynolds number corresponding to the onset of turbulence increases slowly as the values of the flow behavior index n' decrease.

Mixing problems can be challenging in long tubular reactors circulated with an air-lift system and operating with very dense cultures of filamentous microorganisms such as *Spirulina*. Local concentration gradients may be formed, their duration being dependent on the rate of mixing time. These gradients may have deleterious effects on the growth of the organism.

An increase in yield of a *Chlorella* culture was observed when the flow pattern was modified from a laminar one to a turbulent flow (Pirt et al., 1983). This was most likely a result of an improved light regime experienced by the cells (Richmond and Vonshak, 1978) and better mass transfer conditions (Grobbelaar, 1989). The question still to be answered is whether a further increase in culture speed in the turbulent region will enhance the productivity in dense cultures of *Spirulina*. Preliminary experiments performed in tubular bioreactors in the author's laboratory with *Spirulina* cultures grown at different biomass concentrations and recycled at various Reynolds numbers (all within the turbulent region of the flow) have shown no significant differences in the respective productivities (Torzillo et al., 1989, unpublished results).

The Circulating Device and Power Requirement

Another essential aspect for the successful design of a bioreactor is the type of device used for pumping and recycling. The shearing effect of the circulating device used will often dictate whether shear-sensitive and fragile cells, such as mobile or filamentous microalgae, can be cultivated in the particular bioreactor system. The physiological parameters of *Chlorella* cultures grown in a tubular loop reactor and recycled either by a centrifugal pump, a rotary positive displacement pump, or a peristaltic pump, were compared (Pirt et al., 1983). The maximum specific growth rate of the *Chlorella* cultures recycled either by a centrifugal or a rotary positive displacement pump was lower than that in cultures recycled by a peristaltic pump. The adverse effects on the cells caused by centrifugal or rotary positive displacement pumps were proportional to the rotation speed of the pump. A similar conclusion with a tubular system in which circulation was realized by a screw-pump which was found to cause cell damage of *Porphyridium*, was also reported (Gudin and Chaumont, 1991). Both peristaltic and membrane pumps have been tested for many years in the author's laboratory for outdoor culture of *Spirulina*. Cultures can be circulated at a speed of 0.3 m s^{-1} without any evident damage to the trichomes. However, when the culture speed was increased to 0.8 m s^{-1}, the growth yield was reduced by 16 per cent. Microscopic observation of the culture showed a significant increase in the number of trichome fragments (<7 μm in length) in the culture circulated at higher speed, as a result of the increased frequency of their passage through the pump (Torzillo et al., 1993). On the other hand, the use of a screw-pump for the culture recycle was found to produce high shear causing trichome breakage (Torzillo, unpublished results). As a result, air-lift systems seem to be more suitable for the culture recycling of *Spirulina*. The main advantages of air-lift systems are their low shear and relative simplicity of construction. Pirt et al. (1981) found that the maximum biomass output rate of a *Chlorella* culture obtained by using an air-lift reactor exceeded that achieved with a peristaltic pump recycle at the same Reynolds number of 2400. Air-lift tubular reactors have been tested for outdoor culture of *Spirulina* (Torzillo et al., 1993; Richmond et al., 1993; Robinson, 1993). The relation between the air flow rate and the liquid flow rate was investigated in the author's laboratory with a 140 m long tubular loop (i.d. = 2.6 cm). The experiments were performed at three different gas riser heights: 1.65, 2.65 and 3.65 m. The water flow rates were measured in a 1.3 to 28.2 1 min^{-1} range of air flow rates. These experiments showed that the highest water recycle was obtained with the 3.65 m riser (Figure 6.6). However, a Reynolds number > 2100 was reached even with a 1.65 m

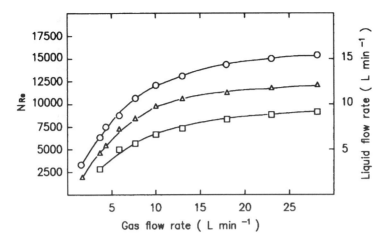

Figure 6.6 Influence of gas flow rate on Reynolds number (N_{Re}) and liquid flow rate in a 140 m long air-lift tubular bioreactor (i.d. = 2.6 cm). The heights of risers (i.d. = 2.6 cm) were: (O) 3.65 m, (Δ) 2.65 m, and (□), 1.65 m. (From Torzillo et al., 1993, with permission.)

gas riser and a flow rate inferior to 2.5 l min^{-1}. Other advantages of air-lift systems are that they have no moving parts in the liquid and are able to achieve high CO_2 transfer rates and oxygen degassing of the culture. It must be pointed out that even air bubbling can produce hydrodynamic stress in the steps involved in passing the gas through the liquid; that is, bubble formation and break up at the orifice of the sparger, bubble coalescence, and bubble bursting at the culture surface when they leave the liquid (Silva et al., 1987). Moreover, considering the design costs, long loops would be advantageous, but they would require a higher air pressure and consequently a switch over from the more efficient rotary air-blower to a more costly and less suitable piston air-compressor. The sum of these considerations is that any mixing system will produce some stressing effect on algal cells; the effect becomes evident when the cultures are circulated at high speeds. Thus, a positive balance must be achieved between mixing that creates proper light and mass transfer regimes and mixing that results in mechanical cell damage. There is no doubt that the design of an efficient circulating device represents a prerequisite to increased productivity and decreased energy input.

The water head required to obtain a given flow rate in tubular reactors is calculated with the equation:

$$h = \frac{V^2}{2g} + \frac{V^2 sL}{2gD} + \frac{V^2 Nb}{2g} \tag{6.3}$$

where V is velocity of flow (m s^{-1}), g is gravitational acceleration (9.81 m s^{-2}), L is length of tubing (m), D is diameter of tubing (m), s is shear factor (dimensionless), N is number of bends in the loop (dimensionless) and b is the bend factor (dimensionless). Tubular reactors have been subjected to a detailed analysis by Weissman et al. (1988). The results obtained for mixing power requirements were similar to the author's calculations. The water head requirements for water recycle in a tubular reactor made of Plexiglas® tubes of 2.6 cm i.d. was reported by the author and co-workers (Torzillo et al., 1993) (Figure 6.7). They observed that as the flow

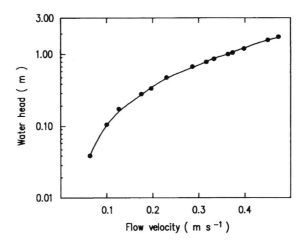

Figure 6.7 Water head required for water recycle in a 140 m long tubular bioreactor, i.d. = 2.6 cm. (From Torzillo et al., 1993, with permission.)

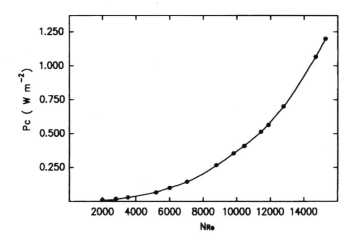

Figure 6.8 Power required per square meter of tubular bioreactor (*Pc*) for water recycle at different Reynolds numbers (N_{Re}). (From Torzillo et al., 1993, with permission.)

velocity increased over 0.1 m s^{-1}, a higher water head was required to obtain a given increase in flow velocity. The power (*Pc*) required for water recycle in the tubular reactor is the product of volumetric flow rate, specific weight and head loss. It is calculated according to the equation:

$$Pc = 9.81 \times 10^3 \, h f \qquad (6.4)$$

where h is water head (m) and f is water flow rate ($\text{m}^3 \text{ s}^{-1}$). The power required per square meter of the area (*Pc/A*) covered by a tubular reactor made with tubes of 2.6 cm to achieve full turbulent flow (Reynolds number > 4000) was $3.93 \times 10^{-2} \text{ W m}^{-2}$ (Figure 6.8). However, according to our calculation, due to culture viscosity, the power requirement to circulate a *Spirulina* suspension of about 3 g l^{-1} at the same Reynolds number would be 0.434 W m^{-2}, i.e. a value more that 10 times higher than that calculated for water.

Limitation of Tubular Bioreactors

Although closed tubular bioreactors seem to provide a promising new technique for mass production of *Spirulina*, information concerning large-scale units is scanty. Overheating of the culture seems to be the major limitation to the use of tubular bioreactors. Indeed, in summer days, the culture temperature can reach 10 to 15 °C over the optimum for *Spirulina* for a few hours. The cost of a cooling device may reduce the economic advantage of closed systems. We have tested three cooling systems for the mass cultivation of *Spirulina* (Torzillo et al., 1986), as follows.

1 Shading of the tubular bioreactor with dark-colored plastic sheets. For an effective control of the culture temperature it was necessary to cover about 80 per cent of the illuminated surface for 5–6 h daily. This caused a strong reduction in the amount of solar radiation received by the culture and consequently in biomass yield.

2 Overlapping the tubes in the north–south orientation in order to reduce the amount of light received by the culture. This system was difficult to install and inadequate for effective control of the culture temperature.

3 Cooling the culture by spraying water on the surface of the bioreactor. The system was operated when the culture temperature reached a critical value of 35 °C and was used about 40 days per year. This cooling device worked very efficiently. However, in order to reduce the amount of water required for cooling, a thermotolerant strain of *Spirulina* (*Spirulina platensis* M2) was selected. This strain was able to grow at up to 42 °C and to tolerate up to 46 °C for a few hours.

Another factor that constrains scaling-up of tubular bioreactors is oxygen accumulation. As aforesaid, with actively growing *Spirulina* cultures circulated at 0.3 m s^{-1}, oxygen degassing stations are required every 100 m of tube (i.d. 2.6–3.0 cm). In order to contain the number of degassing stations, it is necessary to increase the culture speed and/or to use nitrogen for effective oxygen degassing. Otherwise, larger diameter tubes should be used. The adoption of an array of parallel tubes connected by manifolds could also be used to reduce the number of degassing stations (Richmond et al., 1993).

The selection of *Spirulina* strains able to withstand high oxygen concentrations and temperatures could greatly facilitate the bioreactor design and operation and, consequently, reduce the production costs. It is beyond doubt that the tubular bioreactor represents a full controllable culture system that is suitable for scaling-up to industrial scale units that will allow the production of high quality algal biomass.

Acknowledgements

The author wishes to express gratitude to his colleagues at the Centro di Studio dei Microrganismi Autotrofi, Florence, with whom he has been working on outdoor culture of *Spirulina* for many years. The author is grateful to Prof. R. Materassi for fruitful criticism of the manuscript, and to the late Prof. Florenzano for having driven him towards mass culture of *Spirulina* in tubular bioreactors. This research was supported by the National Research Council of Italy, Special Project RAISA, Sub-project No. 4, paper No. 1783.

Spirulina platensis (Arthrospira)

References

ANONYMOUS (1953) Pilot-plant studies in the production of *Chlorella*. In BURLEW, J. S. (Ed.) *Algal Culture: From Laboratory to Pilot Plant*, Chap. 17, Washington, DC: The Carnegie Institution, Publ. No. 600.

BOCCI, F., TORZILLO, G., VINCENZINI, M. and MATERASSI, R. (1987) Growth physiology of *Spirulina platensis* in tubular photobioreactor under solar light. In STADLER, T., MOLLION, J., VERDUS, M. C., KARAMANOS, Y., MORVAN, H. and CHRISTIAEN (Eds), *Algal Biotechnology*, London: Elsevier Applied Science, p. 219.

BOROWITZKA, L. J. and BOROWITZKA, M. A. (1989) Industrial production: methods and economics. In CRESSWELL, R. C., REES, T. A. V. and SHAH, N. (Eds) *Algal and Cyanobacteria Biotechnology*, Chap. 10, New York: Longman/Wiley.

DAVIS, E. A., DETRICK, J., FRENCH, C. S., MILNER, H. W., MYERS, J., SMITH, J .H. C. and SPOEHR, H. A. (1953) Laboratory experiments on *Chlorella* culture at the Carnegie Institution of Washington, Department of Plant Biology. In BURLEW, J. S. (Ed.) *Algal Culture: From Laboratory to Pilot Plant*, Chap. 9, Washington, DC: The Carnegie Institution, Publ. No. 600.

DODGE, D. W. and METZNER, A. B. (1959) Turbulent flow in non-Newtonian systems, *AIChE. J.*, **5**, 189.

FISHER, A. W. (1956) Engineering for algae culture. In *Proc. World Symposium on Applied Solar Energy*, Phoenix, Arizona, sponsored by The Association for Applied Solar Energy, Stanford Research Institute, University of Arizona, Published and Distributed by Stanford Research Institute, Menlo Park, California, p. 243.

GROBBELAAR, J. U. (1989) Do light/dark cycles of medium frequency enhance phytoplankton productivity?, *J. Appl. Phycol.* **1**, 33.

GUDIN, C. and CHAUMONT, D. (1983) Solar biotechnology study and development of tubular solar receptors for controlled production of photosynthetic cellular biomass for methane production and specific exocellular biomass. In PALZ. W. and PIRRWITZ, D. (Eds), *Energy from Biomass*, Dordrecht: D. Reidel, p. 184.

GUDIN, C. and CHAUMONT, D. (1991) Cell fragility – The key problem of microalgae mass production in closed photobioreactor, *Bioresource Technol.*, **38**, 145.

HOSHINO, K., HAMOCHI, M., MITSUHASHI, S. and TANISHITA, K. (1991) Measurements of oxygen production rate in flowing *Spirulina* suspension, *Appl. Microbiol. Biotechnol.*, **35**, 89.

LEE, Y. K. (1986) Enclosed bioreactors for the mass cultivation of photosynthetic microorganisms: the future trend, *Trends in Biotechnol.*, **4**, 186.

LEE, Y. K. and LOW, C. S. (1991) Effect of photobioreactor inclination on the biomass productivity of an outdoor algal culture, *Biotechnol. Bioeng.*, **38**, 995.

METZNER, A. B. and REED, J. C. (1955) Flow of non-Newtonian fluids – correlation of the laminar, transition, and turbulent-flow regions, *AIChE. J.*, **1**, 434.

PIRT, S. J., BALYUZI, M. J., LEE, Y. K., PIRT, M. W. and WALACH, M. R. (1981) A photobioreactor to store solar energy by microbial photosynthesis. In *Brochure of Queen Elizabeth College, University of London*, p. 33.

PIRT, S. J., LEE, Y. K., WALACH, M. R., PIRT, M. W., BALYUZI, H. H. M. and BAZIN, M. J. (1983) A tubular bioreactor for photosynthetic production of biomass from carbon dioxide: design and performance, *J. Chem. Biotechnol.*, **33 B**, 35.

RICHMOND, A. (1990) Large scale microalgal culture and applications. In ROUND, F. F. and CHAPMAN, D. J. (Eds) *Progress in Phycological Research*, Vol. 7, Chap. 4, Bristol: Biopress.

RICHMOND, A. and VONSHAK, A. (1978) *Spirulina* culture in Israel, *Arch. Hydrobiol. Beih. Engebn. Limnol.*, **11**, 274.

RICHMOND, A., BOUSSIBA, S., VONSHAK, A. and KOPEL, R. (1993) A new tubular reactor for mass production of microalgae outdoors, *J. Appl. Phycol.*, **5**, 327.

114

ROBINSON, L. F. (1993) Biotechna Graesser A. P. Ltd., Personal communication.

SILVA, H. J., CORTINAS, T. and ERTOLA, R. J. (1987) Effect of hydrodynamic stress on *Dunaliella* Growth, *J. Chem. Tech. Biotechnol.*, **40**, 41.

SKELLAND, A. H. P. (1967) Non-Newtonian flow and heat transfer, New York: Wiley.

TAMIYA, H. (1956) Growing *Chlorella* for food and feed. In *Proc. World Symposium on Applied Solar Energy*, Phoenix, Arizona, p. 231, Menlo Park, California: Stanford Research Institute.

TAMIYA, H., HASE, E., SHIBATA, K., MITUYA, A., IWAMURA, T., NIHEI, T. and SASA, T. (1953) Kinetics of growth of *Chlorella*, with special reference to its dependence on quantity of available light and on temperature. In Burlew, J. S. (Ed.) *Algal Culture: From Laboratory to Pilot Plant*, Chap. 16, Washington, DC: The Carnegie Institution, Publ. No. 600.

TORZILLO, G., GIOVANETTI, L., BOCCI, F. and MATERASSI, R. (1984) Effect of oxygen concentration on the protein content of *Spirulina* biomass, *Biotechnol. Bioeng.*, **26**, 1134.

TORZILLO, G., PUSHPARAJ, B., BOCCI, F., BALLONI, W., MATERASSI, R. and FLORENZANO, G. (1986) Production of *Spirulina* biomass in closed photobioreactors, *Biomass*, **11**, 61.

TORZILLO, G., SACCHI, A., MATERASSI, R. and RICHMOND, A. (1991) Effect of temperature on yield and night biomass in *Spirulina platensis* grown outdoors in tubular photobioreactors, *J. Appl. Phycol.*, **3** 103.

TORZILLO, G., CARLOZZI, P., PUSHPARAJ, B., MONTAINI, E. and MATERASSI, R. (1993) A two-plane tubular photobioreactor for outdoor culture of *Spirulina*, *Biotechnol. Bioeng.*, **42**, 891.

TREDICI, M. R., ZITTELLI, G. C. and BIAGIOLINI, S. (1992) Influence of turbulence and areal density on the productivity of *Spirulina platensis* grown outdoors in a vertical alveolar panel. In *Proc. 1st European Workshop on Microalgal Biotechnology, Algology*, p. 58, Potsdam-Rehbrücke: Institut für Getreideverarbeitung.

WEISSMAN, J. C., GOEBEL, R. P. and BENEMANN, J. R. (1988) Photobioreactor design: mixing, carbon utilization, and oxygen accumulation, *Biotechnol. Bioeng.*, **31**, 336.

7

Cultivation of *Spirulina (Arthrospira) platensis* in Flat Plate Reactors

MARIO R. TREDICI AND GRAZIELLA CHINI ZITTELLI

Historical Background

Today, the commercial production of *Spirulina* biomass is carried out exclusively in open systems (Vonshak and Richmond, 1988). Closed reactors have so far been employed for research and in small field installations (Torzillo et al., 1986, 1993; Richmond et al., 1993; Chini Zittelli et al., 1993).

Flat reactors or flat culture chambers have been used to grow photosynthetic microorganisms in the laboratory. They provide a simple geometry and greatly facilitate the measurement of irradiance on the culture surface. Despite their apparent simplicity, however, they have not been used for commercial mass cultivation of algae.

The first flat culture unit devised for mass production of algae was the rocking tray used by Milner (1953) in the early work on algal mass cultivation to grow dense *Chlorella* cultures in a thin turbulent layer. This culture system consisted of a 2.65 m^2, 9 cm-deep tray covered with transparent wire-reinforced plastic and supported on a shaker. Both the frequency and the amplitude of the rocking were adjustable. The daily yield varied with the weather. At a mean culture depth of 1.7 cm, an average productivity of 8.2 g m^{-2} day^{-1} was obtained in a 20-day run. An improved version of this type of culture unit was used by Soeder et al. (1981) for the axenic synchronous cultivation of *Scenedesmus obliquus* under laboratory conditions. Both reactors are horizontal two-phase systems. A different approach was followed by Anderson and Eakin (1985) who cultivated *Porphyridium cruentum* outdoors for polysaccharide production using 3 m^2, 5–20 cm-deep flat plates (made of stainless steel and covered with glass or plastic) which were placed at an inclination from the horizontal to maximize solar irradiance interception. About 20 g of polysaccharide m^{-2} day^{-1} were produced during the summer when temperature control was provided.

Essentially, no other report about flat reactors used for mass production of algae can be found with the exclusion of alveolar panels. (Covered troughs or ponds, which essentially relate to open systems except for the transparent cover, are not addressed here.) These latter systems are constructed from transparent polyvinyl chloride, polycarbonate or polymethyl methacrylate sheets which are internally partitioned to form rectangular channels called 'alveoli' and are commercially available in standard thicknesses from 4.5 to 40 mm. Their introduction in algae cultivation is relatively recent, it being only in the mid-1980s that they were independently used in France and Italy. It seems that their adoption has been hampered by some prejudice regarding deficiencies in culture flow control and the engineering problems of suitable cost-effective panels (Pirt et al., 1983). The use of alveolar plates will probably spread in the near future since they allow a cheap and strong flat reactor to be built with a minimum of work and without any need to reinforce the walls so that they may withstand deformation due to water pressure (see Hase and Morimura (1971) for an example of a flat culture chamber made of reinforced non-alveolar acrylate sheets).

The first use of alveolar plates as culture systems for algae was carried out at the Centre d'Etudes Nucléaires de Grenoble (France), where Ramos de Ortega and Roux (1986) experimented with 6 m-long, 0.25 m-wide, 40 mm-deep, triple wall (double row of channels) panels for growing *Chlorella* in a greenhouse and outdoors. The plates were laid horizontally on the ground, the upper layer of channels being used for algal growth, the lower for thermoregulation. The culture suspension was circulated through a pump. A productivity of $23-24$ g m^{-2} day^{-1} was achieved in the summer. The authors stressed the fact that the double layer plate permitted an efficient thermoregulation of the culture and that significantly higher performances were obtained with the panels compared with two other closed systems: flexible sheets and semirigid tubes, all of them made of polyvinyl chloride.

In 1988, the Consiglio Nazionale delle Ricerche (CNR) of Italy presented a flat alveolar reactor for algae cultivation at the exhibition 'ITALIA 2000' held in Moscow. The panel was placed vertically, with channels (and hence culture flow) running parallel to the ground. A pump was used to circulate the suspension. In the same year, a different design was patented by CNR (Tredici et al., 1988). In this model, the plates were placed vertically, but presented two main modifications with respect to the previous design: the channels were perpendicular to the ground, and mixing of the culture was achieved through bubbling air at the bottom of the reactor. This latter reactor type (Figure 7.1A,B) has been used extensively by the Florence group to cultivate cyanobacteria and microalgae outdoors (Chini Zittelli et al., 1993; Tredici et al., 1991, 1992, 1993; Tredici and Materassi, 1992).

More recently, at the Institut für Getreideverarbeitung (Potsdam-Rehbrucke, Germany), several cultivation units based on 16 and 32 mm-thick alveolar panels have been devised by Pulz and co-workers (Pulz, 1992). The plates are placed vertically, with channels arranged horizontally, just 20 cm apart, to form a compact structure (Figure 7.1C). The culture is recirculated by a piston pump. Units from 100 to 10 000 l have been developed and tested, both indoors and outdoors, with different microalgae and cyanobacteria, including *S. platensis*. Productivity levels of 1.3 g l^{-1} day^{-1}, corresponding to 28 g m^{-2} day^{-1} on a basis of illuminated surface area and to 120 g day^{-1} per square meter of occupied land, have been obtained under natural illumination. The main problems that have emerged are related to build-up of oxygen (up to $30-40$ mg l^{-1} at the outlet of the circuit) and fouling.

Muller-Feuga (1993) at Cadarache Centre of CEA (Commisariat à l'Energie Atomique, Saint Paul les Durance, France) has developed and operated a 4 m^2 alveolar photobioreactor to grow axenic cultures of *Haematococcus pluvialis* and

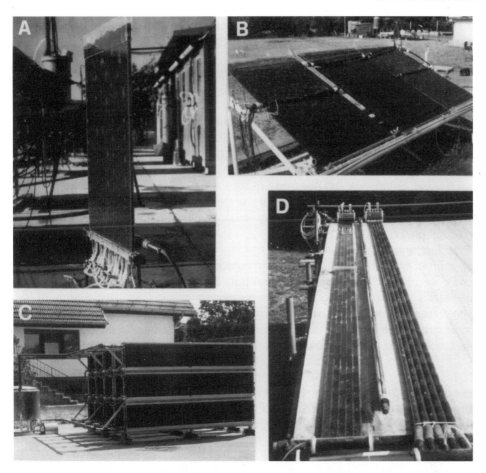

Figure 7.1 Alveolar plates employed to grow *Spirulina* spp. outdoors: (A) 0.3 m² vertical bubble column plate (Florence, Italy); (B) 2.2 m² inclined plates (Florence, Italy); (C) vertical pump mixed plates (Postdam-Rehbrucke, Germany, Courtesy of Dr O. Pulz); (D) 1.3 m² reactors used in Florence to compare the performance of bubble column plates (left) and bubble column tubular systems (right).

Porphyridium cruentum outdoors. The reactor is made of four 16 mm-thick alveolar plates which are connected in series and placed facing south on a 45°-inclined support. Temperature control is provided through a second plate, in close contact to the culture panel surface, into which cooling (in the summer) or heating (in the winter) water is circulated. For cleaning, a small rectangular plug is forced up and down each channel by the pressure of the culture flow. The culture is circulated by a stainless steel flexible impeller pump, and the circuit is completely closed and slightly pressurized to avoid contamination. The system, computer-controlled, has been designed to produce algal biomass of high value and pharmaceuticals. No productivity data are available at this time.

A 0.5 m² flat plate air-lift reactor has been used at the Department of Biochemical Sciences of the Scottish Agricultural College (UK) to grow algal cultures at high biomass concentration under artificial light (Ratchford and Fallowfield, 1992). Among the advantages of the alveolar plates, the authors cite the more uniform distribution of irradiance, the attainment of high cell concentration at the stationary phase, and the fact that, because of the structural strength of the sheet, less support is required compared with flexible plastic tubes.

The plates experimented with at Florence are constructed from 16 or 32 mm thick Plexiglas® (polymethyl methacrylate) or Makrolon® (polycarbonate) alveolar sheets commonly available on the market at a cost of US$ 20–30 per square meter. A more detailed description of these systems can be found in Tredici et al. (1991) and Tredici and Materassi (1992). The main features which differentiate the reactors developed at Florence from the others are the vertical or tilted inclination from the horizontal of the channels and the fact that no mechanical device is used to mix the cell suspension, circulation and degassing of the culture being attained through bubbling air at the bottom of the panel (typically at the base of each channel). Hence, these systems can be classified as 'bubble column' flat photobioreactors or 'bubble column' plates.

In the following sections, more specific information on the performance of *Spirulina (Arthrospira) platensis* grown outdoors in elevated 'bubble column' plates under the climatic conditions of Central Italy (Florence) is given, with reference to the effect of areal density, turbulence, oxygen tension and inclination of the photobioreactor on productivity and biomass composition. Comparison is made with other culture systems used to grow *Spirulina* spp. in order to let the reader evaluate the potential and the drawbacks of this new photobioreactor design.

Productivity of *S. platensis* Grown in Different Cultivation Systems Under the Climatic Conditions of Central Italy (Florence)

One of the current major handicaps that hamper the commercial exploitation of *Spirulina* and other microalgae is the low yield per unit of cultured surface area (Richmond, 1990). *Spirulina* cultures in small, well-managed, raceway ponds can attain productivities of 15 to 19 $g m^{-2} day^{-1}$ (Richmond, 1990; Materassi et al., 1980) and less than half that figure is attained in commercial production plants (Vonshak and Richmond, 1988). In closed tubular reactors, productivities of 25–28 $g m^{-2} day^{-1}$ have been achieved (Torzillo et al., 1986, 1993; Chini Zittelli et al., 1993). These yields have been obtained, however, in small units operating for a few weeks or months under strictly controlled conditions.

Table 7.1 shows the areal and volumetric productivities of *S. platensis* cultivated in different systems under the climatic conditions of Florence. As can be seen, air-lift and bubble column tubular reactors achieve about 90 per cent higher areal yields than ponds and 15 per cent higher areal yields than elevated plates. It should be pointed out, however, that plates, because of their high surface-to-volume ratio (80 m^{-1}), attain by far the highest volumetric yield (2.1 $g l^{-1} day^{-1}$) and allow operations at very high cell concentration (5 $g l^{-1}$). Under certain conditions, these features will influence the choice of the culture system more than will the areal yield. Table 7.2 shows typical areal productivities achieved in Florence with *S. platensis* in elevated plates during different seasons of the year. It is worth noting that, besides a good performance during the summer and autumn, because of the elevated arrangement of the reactor, significant areal outputs (9 $g m^{-2} day^{-1}$) are attained even in winter (in cultures artificially heated) when very low solar energy is available for algal growth.

120

Table 7.1 Productivity of *S. platensis* grown outdoors in different cultivation systems under the climatic conditions of Florence (summer)

Reactor type	Surface to volume ratio (m^{-1})	Mean algal concentration (g l^{-1})	Volumetric productivity (g l^{-1}d^{-1})	Areal productivity (g m^{-2}d^{-1})	Reference
Raceway ponds	10	0.8	0.15	15.0	Materassi et al., 1980
Tubular 'pump mixed' reactor	10	0.7	0.25	25.0	Torzillo et al., 1986
Tubular 'air lift' reactor	54	4.2	1.50	27.8	Torzillo et al., 1993
Tubular 'bubble column' reactor	44	3.7	1.30	28.1	Chini Zittelli et al., 1993
Elevated panels	80	5.0	2.10	24.3	Tredici et al., 1993

Data for elevated panels refer to the reactor illuminated surface area; for the other systems, to the occupied land surface area.

Table 7.2 Productivity of *S. platensis* grown in elevated plates during different seasons under the climatic conditions of Florence (Tredici et al., 1991, 1993)

Season	Angle of inclination (°)	Areal productivity[a] (g m^{-2}d^{-1})	Mean solar irradiance[b] (MJ m^{-2}d^{-1})
Summer	22–33	24.3	25.5
	90	17.4	25.5
Autumn	25–30	15.3	17.3
Winter	57–67	9.0	7.6

[a] Refers to the illuminated surface of the reactor.
[b] Refers to the horizontal surface.

Outdoor Cultivation of *S. platensis* in Elevated Bubble Column Plates

Influence of Areal Density

The importance of maintaining an optimal areal density (OAD) (i.e. a standing crop per unit of surface area which will result in the highest yield) in outdoor algal cultures has been well documented (Materassi et al., 1980; Richmond, 1990; Richmond and Vonshak, 1978; Richmond and Grobbelaar, 1986; Vonshak et al., 1982; Hartig et al., 1988). The highest output rates with *S. platensis* grown outdoors in open ponds have been achieved at areal densities between 52 and 75 g m^{-2} (Vonshak et al., 1982). With the same organism, it has been found that the areal density needed to obtain maximum productivity changes with mixing rate and depth of the culture: the OAD is 90 g m^{-2} with a depth of 15 cm, and 60 g m^{-2} in a culture 7.5 cm-deep (Richmond and Grobbelaar, 1986). *S. platensis* cultivated in tubular reactors of 13 cm internal diameter in the summer attained about 35 per cent higher

121

productivity at an areal density of 60 g m^{-2} as opposed to 120 g m^{-2} (Torzillo et al., 1986). The influence of the population density on productivity decreased as autumn approached, probably because of a greater limitation exerted by the temperature. The reduction of productivity at areal density values higher than the OAD was less evident in tubular reactors of 2.6 cm bore (Torzillo et al., 1993).

The effect of areal density on productivity of *S. platensis* grown outdoors in elevated plates was investigated by Tredici et al. (1992) (Table 7.3). The highest gross productivity (total amount of biomass synthesized during the daylight hours per unit of surface area) was achieved at mean initial areal densities of 45.8 and 53.8 g m^{-2}. At 53.8 g m^{-2} (as at 68.9 g m^{-2}), however, dark respiration curtailed the daytime productivity of a much higher percentage, thus significantly decreasing the net productivity (gross productivity less the amount of biomass lost because of respiration during the night). The highest net productivity (24.3 g m^{-2} day^{-1}) was achieved at an initial areal density of 45.8 g m^{-2}, i.e. at an OAD value slightly lower than that found for open ponds and tubular reactors. At an areal density of 22.6 g m^{-2} the culture collapsed after about one week. An explanation for this phenomenon, which was observed in three independent experiments carried out at population densities lower than 30 g m^{-2}, is not given. Very probably it might have been caused by the combined action of high photosynthetic photon flux densities (HPPFD) impinging on diluted cultures and high pO$_2$ (35 mg of dissolved oxygen l^{-1} were measured in these latter cultures at midday), since, under laboratory conditions, *S. platensis* grown at low areal densities achieved a stable, high productivity even under a HPPFD of 2000–2200 μmol photon m^{-2} s^{-1} (typical values of outdoor conditions at noontime), when the oxygen tension was maintained below 20 mg l^{-1} (Tredici et al., 1992).

Influence of Oxygen Tension and Turbulence

Dissolved oxygen concentrations over 35 mg l^{-1} may be easily reached in outdoor dense algal cultures at midday. Such high oxygen levels are toxic to most algae and,

Table 7.3 Influence of areal density on the productivity of *S. platensis* grown outdoors in elevated plates (Tredici et al., 1992)

Mean initial areal density[a] (g m^{-2})	Areal productivity (g m^{-2} d^{-1})		NBL (%)
	Gross	Net	
22.6	([b])		([b])
45.8	29.7	24.3	18.2
53.8[c]	30.8	21.6	29.9
68.9	26.4	18.5	29.9

Plates were inclined at 33°.
[a] Measured in the early morning after dilution.
[b] The culture doesn't reach the steady state at this areal density.
[c] This experiment was carried out using a 2.5 m^2, 3.2 cm-thick panel.
NBL: Night biomass loss calculated as percentage of the daylight productivity.

if coupled with prolonged exposure to full sunlight, may lead to photooxidation and photooxidative death of the culture (Richmond, 1986). Accumulation of photo-synthetically generated oxygen becomes a particularly serious problem in closed, high surface-to-volume ratio photobioreactors operated outdoors. At maximal rates of photosynthesis, such as those observed at midday, algal cultures in 1 cm tubes mixed at $20 \, cm \, s^{-1}$ may accumulate about 100 mg of oxygen l^{-1} after 100 m, i.e. over 10 times the oxygen content of air-saturated water (Weissman et al., 1988). In tubes of greater diameter the situation is less dramatic, but even in 14-cm bore tubular reactors, dissolved oxygen may reach levels that significantly reduce the productivity and protein content of *Spirulina* cultures (Torzillo et al., 1986).

Because of their high surface-to-volume ratio, alveolar plates may attain, at midday, volumetric biomass output rates higher than $400 \, mg \, l^{-1} h^{-1}$ (Tredici and Materassi, 1992) which, assuming a photosynthetic quotient of one, lead to oxygen evolution rates of about $9 \, mg \, l^{-1} min^{-1}$. Very efficient degassing is required under these conditions so as not to endanger the well-being of the culture. Air-bubbling at the base of each channel has proved to be an efficient technique in order to prevent the oxygen from building up to dangerous levels in these culture systems (Tredici et al., 1991). Besides, compared with tubular serpentines, in which oxygen accumulates during the loop cycle and is removed only at the degassing station or in the gas riser, bubble column plates present the inherent advantage of allowing the simultaneous degassing of the whole cultural suspension, and, consequently, no large gradients of dissolved oxygen are experienced by the cultured cells.

The beneficial effects that stirring has on growth and productivity of algal mass cultures are well known (Richmond and Vonshak, 1978; Richmond, 1990). Turbu-lence prevents cells from settling, breaks down diffusion barriers around the cells and, by forcing the algal cells to experience alternating periods of light and darkness, enhances photosynthetic efficiency. Conflicting results have been reported, however, for light–dark fluctuations of medium frequency (i.e. in the order of seconds to minutes) that prevail in mixed mass cultures as regards their stimulatory effect on productivity, and the true role of the mixing rate in determining algal productivity still remains to be clarified (Richmond, 1990; Weissman et al., 1988; Grobbelaar, 1991). Positive effects of intense stirring on productivity have been observed only in outdoor algal cultures grown in paddle-wheel mixed raceway ponds (Richmond and Vonshak, 1978; Laws et al., 1983). No data are available, to our knowledge, for closed reactors. The reason for this lack of information may be that experiments carried out to demonstrate the positive effect of high stirring rates in tubular reactors have made use of pumps as mixing devices. Hence, the existence of a positive effect, if any, may have been masked by a reduced yield due to mechanical cell damage (Torzillo et al., 1993; Gudin and Chaumont, 1991).

In bubble column photoreactors, the supply of air serves the double purpose of providing turbulence and degassing (oxygen removal). Hence, it is rather difficult to evaluate whether the influence of the chosen air-bubbling rate on productivity arises from the degree of turbulence achieved or from the resulting dissolved oxygen content at the equilibrium. The influence of the rate of air supply on productivity and biochemical composition of *S. platensis* grown in bubble column plates was thoroughly studied by Tredici et al. (1992, 1993). The beneficial effects of a high airflow rate were evident. At $1.21 \, 1 \, l^{-1} min^{-1}$ (litres of air bubbled per litre of cell suspension per minute), as opposed to $0.36 \, 1 \, l^{-1} min^{-1}$, gross and net productivity increased by 28 and 35 per cent, respectively, and the pO_2 stabilized at significantly lower values (Table 7.4, experiment 1). The question arose whether the observed positive influence on productivity of the higher airflow rate was due to a more adequate light regimen resulting from the higher turbulence, or whether it was a consequence of the reduced oxygen tension. The first hypothesis seemed to be somewhat supported by the observation that the culture grown at the higher airflow rate showed a reduced phycobiliprotein and chlorophyll *a* content, thus indicating a condition of higher light availability to the single cell (Wyman and Fay, 1987). It cannot be completely ruled out, however, that the lower pigment content of the cells grown at the higher airflow rate may have resulted, at least partially, from a lower areal density (the two experimental cultures had, in fact, the same initial cell concentration, but differed as to gas hold-up, hence they had different areal densities). It must be considered, however, that, while the initial (early morning) areal density varied between the two cultures by 5.9 $g\,m^{-2}$, the mean daytime areal density varied by less than 1 $g\,m^{-2}$. According to our experience, this small difference cannot give rise to a measurable change in the cell pigment content. On the other hand, an effect of the decreased dissolved oxygen concentration must be excluded, since this condition would, on the contrary, favor phycobiliprotein synthesis (Torzillo et al., 1984). To single out the influence on productivity of the sole turbulence, a second experiment was carried out with cultures operated at two different airflow rates, but maintained at the same dissolved oxygen content (Table 7.4, experiment 2). Also in this case, the higher airflow rate enhanced the culture productivity (by about 15 per cent) and reduced the pigment content of the biomass, thus confirming that a different amount of light was available to the single cell in the two differently bubbled cultures. The influence of pO_2 was assessed in two *S. platensis* cultures bubbled with 1.21 $ll^{-1}\,min^{-1}$. The increase of the dissolved oxygen concentration from 20–22 to 35–36 $mg\,l^{-1}$, which only slightly modified the cell pigment content, caused a 12–15 per cent reduction of productivity (Table 7.4, experiment 3). No damage to the cells nor any significant change in carbohydrate or protein content was observed at the higher air bubbling rates. It thus may be suggested that a higher bubbling rate enhances the productivity by providing a more adequate light regimen to the cells, as well as by reducing the oxygen levels in the culture medium.

Table 7.4 Influence of airflow rate and oxygen tension on the productivity and pigment content of *S. platensis* grown outdoors in elevated plates (Tredici et al., 1992)

Experiment	Airflow rate $(l\,l^{-1}\,min^{-1})$	pO_2 $(mg\;l^{-1})$	Productivity $(g\;m^{-2}d^{-1})$		Pigment content $(\%\;dw)$	
			Gross	Net	PBP	Chl. *a*
1	0.36	35–36	23.5	17.4	5.40	1.12
	1.21	20–21	30.2	23.5	4.53	1.05
2	0.36	30–32	23.4	17.5	6.34	1.30
	1.21	30–32	26.8	19.6	5.50	1.17
3	1.21	20–22	29.4	21.7	5.62	1.20
	1.21	35–36	25.9	18.5	5.46	1.11

Inclination of bioreactors was 36° during the first experiment and 41° during the second and third experiment. The values of pO_2 refer to a plateau reached during midday. All areal parameters are on the basis of illuminated surface.
PBP = phycobiliproteins; Chl. *a* = chlorophyll *a*.

Influence of the Inclination of the Photobioreactor

One of the main impediments to the efficient use of solar energy through algae cultivation is the continuous variation of the available solar radiation throughout the day. Solar irradiance is very low in the early morning and in the evening hours, and high enough to oversaturate and even cause photoinhibition at noontime. While the diurnal pattern of the solar irradiance incident on a horizontal reactor (e.g. an open pond) cannot be easily modified, elevated systems (e.g. inclined plates) are capable of being oriented and inclined at various angles to the sun and offer the possibility to vary, continuously or discontinuously, the natural distribution of the solar radiation falling on the reactor surface. A higher photosynthetic efficiency should be expected from cultures grown in these latter systems.

The effect of the photobioreactor orientation or inclination on productivity of outdoor algal cultures has rarely been investigated. Lee and Low (1991) determined the productivity of outdoor cultures of *Chlorella pyrenoidosa* in a 0.8 m^2 tubular-loop photobioreactor inclined at various angles to the horizontal. Productivities between 26 and 30 g m^{-2} day^{-1} were achieved irrespective of the angle of inclination. Because of the particular reactor design (two tubular panels were connected by hinges, erected at an angle to the ground and placed in an east–west direction) and, above all, because of the location of the experimental facilities (ca. 1° latitude north), the findings of these researchers cannot be generalized and in particular they cannot be applied to high-latitude regions.

The effect of the inclination of the photobioreactor on the daily biomass output rate of *S. platensis* was studied by Tredici et al. (1993). A south-facing, sun-oriented panel (whose inclination was changed monthly so as to receive the maximum solar radiation on a daily basis) was compared with a south-facing, vertical one. As expected, the angle of inclination of the photobioreactor significantly influenced the culture productivity. At an optimal areal density of 46–48 g m^{-2}, the net biomass output rate in the 30°-inclined reactor was 40 per cent higher (17.4 and 24.3 g m^{-2} day^{-1} were attained in the vertical and in the 30°-inclined reactor, respectively). It is worth mentioning, however, that the culture in the vertical reactor on a daily basis received less than 60 per cent of the solar radiation incident on the inclined system and was on the whole more efficient. The influence of inclination was maximal at suboptimal areal densities. At an areal density of 31–32 g m^{-2}, the culture in the vertical panel attained a steady net productivity of 13.6 g m^{-2} day^{-1}, while the culture in the 30°-inclined reactor collapsed after about one week (see above for a possible cause of this phenomenon).

The diurnal productivity pattern of cultures grown in two differently inclined plates in the summer is illustrated in Figure 7.2; the hourly productivity of a culture grown in a 30°-inclined reactor follows the pattern of solar irradiance quite closely, reaching a peak at 12:30. On a few days (not considered in drawing the typical curves) the peak of productivity preceded the peak of solar irradiance by 1 to 1.5 h. The diminution of productivity with solar irradiance at its maximum was ascribed to photoinhibition of photosynthesis, although this could not be demonstrated by measurements of the light-limited and light-saturated oxygen evolution capacity of the cultures. The typical productivity patterns of *S. platensis* grown in 30°-inclined plates are in substantial agreement with the diurnal patterns of growth displayed by this cyanobacterium in outdoor raceway ponds in mid-summer (Richmond et al., 1990). The typical diurnal productivity pattern of the culture grown in the vertical system is more difficult to explain. The PPFD impinging on the surface of the vertical reactor is always

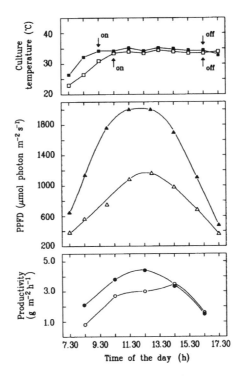

Figure 7.2 Diurnal productivity of *S. platensis* cultivated in 2.2 m² elevated plates. Open symbols: vertical reactor; closed symbols: 30°-inclined reactor. The arrows indicate the start and end of cooling. The air-bubbling rate was 0.8 ll^{-1} min^{-1}.

significantly lower than that on the 30°-inclined one (about 40 per cent less on a daily basis), hence a lower hourly yield should be expected from this system. This is the case, in fact, for the first part of the day (besides, it should be remembered that the culture in this reactor reaches the optimal temperature for growth 1.5 h later), while, from 14:30 onwards, productivity in the two differently inclined reactors is the same, in spite of the very different amount of radiant energy absorbed. The reason for the higher efficiency achieved by cultures in the vertical system during the evening hours is not clear. Richmond et al. (1990) suggested that photoinhibited cells, such as the cells grown in the 30°-inclined plates, seem to respond more sharply to the reduction of light intensity as afternoon progresses. A lower degree of damage caused by phoinhibition could explain the relatively higher efficiency displayed after midday by the cultures grown in the vertical systems.

The effect of temperature on the diurnal productivity pattern of cultures grown in south-facing, 30°-inclined plates was also studied. One of the cultures was artificially heated during the early hours of the morning to quickly increase its temperature to 35 °C. The results of this experiment are summarized in Figure 7.3. The typical productivity pattern is observed, the sole significant difference being a higher biomass output rate at noontime that can be related to the higher airflow rate adopted to mix the culture suspension. The heated culture, which at 7:30 (one and a half hours earlier than the non-heated one) reached the optimum temperature for growth, attained a 10 per cent higher yield on a daily basis and a 69 per cent higher productivity from 7:30 to 9:00 (Figure 7.3). Although the positive influence on

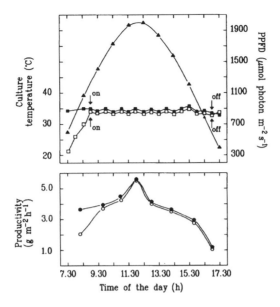

Figure 7.3 Diurnal productivity of *S. platensis* cultivated in 0.3 m², 30°-inclined plates with and without heating during the morning. The diurnal patterns of the photosynthetic photon flux density (PPFD) (triangle) at the surface of the reactors and of the cultures' temperature (squares) are also shown. Open symbols: non-heated reactor; closed symbols: heated reactor. The air-bubbling rate was 2.7 ll⁻¹ min⁻¹.

productivity of elevating the culture temperature unequivocally demonstrates that growth in elevated plates is limited by temperature during the morning, the time course of the diurnal productivity of both the non-heated culture and of that kept at the optimal growth temperature for the whole daylight period shows the existence of a strict correlation between biomass output rate and irradiance. The complex relationships between temperature, solar irradiance and areal density, and their role in regulating algal growth in outdoor culture, have been amply discussed in previous reports (Richmond, 1990; Richmond and Vonshak, 1978; Vonshak et al., 1982).

Prospects

Bubble column plates have proved to be a powerful tool for studying the physiology of outdoor algal mass cultures. These systems allowed us to investigate the influence of some important parameters (e.g. turbulence, accumulation of photosynthetically generated oxygen, inclination of the photobioreactor) on the performance of thin (1 cm) and dense algal suspensions.

Elevated, bubble column plates could be used profitably in the mass cultivation of *Spirulina* spp. and other microalgae. Their limited thickness allows operation at high cell concentrations (typically in the range of 4 to 6 gl⁻¹) and high volumetric productivities (higher than 2 gl⁻¹day⁻¹) to be achieved, with significant technical and economical benefits. Air-bubbling has been shown to be a suitable technique to provide mixing and degassing of concentrated algal suspensions in these systems: while causing no or little damage to the cells, air-bubbling at the base of the reactor

127

ensures high turbulence and maintains low pO_2 levels even during the central daylight hours; in addition, it provides an efficient scouring action which prevents wall growth, an aspect of utmost importance that is often ignored on account of the limited duration of the experiments carried out. The variable inclination of the reactor with respect to the sun's rays offers the possibility to eliminate or at least reduce the harmful effects of high solar radiation falling on the reactor surface at noontime and may enhance the photosynthetic efficiency. Combined with the closed environment and a strict control over some of the main growth parameters (e.g. temperature and pH), these features of the elevated plates ensure efficient protection from contaminants and allow sustained high areal productivities and a biomass of good quality to be attained. Yet, in our opinion, plates suffer from a main limitation that could hamper their future commercial application. Since the single cultivation unit occupies only a few square meters of surface area, the scaling up to commercial dimension of this system will require the assembling of hundreds of units. Although technically complex, this solution has been adopted by the German group that has built impressive facilities in Potsdam-Rehbrucke. A different approach to scaling up of the system has been used in Florence. While alveolar plates are extensively used for research and small-scale production of algal biomass under artificial illumination, for the outdoor mass cultivation of microalgae a new design has been adopted in which several tubes of transparent material are laid parallel on a surface inclined at a very small angle to the horizontal. The tubes are connected through pipes at the top and bottom to construct the typical structure of the bubble column plate (see Figure 7.1D for an example of a small unit of this type). In this way, a single operating unit of 200–300 m^2 surface area could be built. Air/CO_2 is sparged at the bottom of the tubes with bubbles rising slowly along the length of the reactor and providing, as in the plates, gas exchange, mixing and scouring of the inner tube surface. A reactor of this type was tested recently in the outdoor cultivation of *S. platensis*, with outstanding results (Chini Zittelli et al., 1993). Obviously, no single reactor design can be considered optimal for all situations, and both the German approach and our own could be applied in the scaling up of the system.

It will take several years, if ever, before closed systems can replace the low-cost open raceway in the production of raw biomass and commodity products from microalgae. In the case of the production of specialty chemicals or clean biomass for special purposes, the prospects are somewhat different. When high volumetric productivity, low contamination, strict control and high reliability of the process are required, when in other words, more quality than quantity is the target, the choice of closed reactors, either tubular or flat, is mandatory.

Acknowledgement

This research was supported by the National Research Council of Italy, Special Project RAISA, sub-project No. 4, paper No. 1782.

References

ANDERSON, D. B. and EAKIN, D. E. (1985) A process for the production of polysaccharides from microalgae, *Biotechnol. Bioeng. Symp.*, **15**, 533.

CHINI ZITTELLI, G., BIAGIOLINI, S., PINZANI, E. and TREDICI, M. R. (1993) Vertical thin layer photobioreactors: towards the overcoming of their major drawbacks. Presented at *6th Int. Conf. on Applied Algology*, Ceske Budejovice, Czech Republic, 6–11 September.

GROBBELAAR, J. U. (1991) The influence of light/dark cycles in mixed algal cultures on their productivity, *Bioresource Technol.*, **38**, 189.

GUDIN, C. and CHAUMONT, D. (1991) The key problem of microalgae mass production in closed photobioreactors, *Bioresource Technol.*, **38**, 145.

HARTIG, P., GROBBELAAR, J. U., SOEDER, C. J. and GROENEWEG, J. (1988) On the mass culture of microalgae: areal density as an important factor for achieving maximal productivity, *Biomass*, **15**, 211.

HASE, E. and MORIMURA, Y. (1971) Synchronous culture of *Chlorella*. In SAN PIETRO, A. (Ed.) *Methods in Enzymology*, Vol. 23, p. 78, New York: Academic Press.

LAWS, E. A., TERRY, K. L., WICKMAN, J. and CHALUP, M. S. A. (1983) Simple algal production system designed to utilize the flashing light effect, *Biotechnol. Bioeng.*, **25**, 2319.

LEE, Y. K. and LOW, C. S. (1991) Effect of photobioreactor inclination on the biomass productivity of an outdoor algal culture, *Biotechnol. Bioeng.*, **38**, 995.

MATERASSI, R., BALLONI, W., PUSHPARAJ, B., PELOSI, E. and SILI, C. (1980) Coltura massiva di *Spirulina* in sistemi colturali aperti. In MATERASSI, R. (Ed.) *Prospettive della Coltura di Spirulina in Italia*, p. 241, Roma: Consiglio Nazionale delle Ricerche.

MILNER, H. W. (1953) Rocking tray. In BURLEW, J. S. (Ed.) *Algal Culture: From Laboratory to Pilot Plant*, Publ. No. 600, p. 108, Washington, DC: The Carnegie Institution.

MULLER-FEUGA, A. (1993) Personal communication.

PIRT, S. J., LEE, Y. K., WALACH, M. R., PIRT, M. W., BALYUZI, H. H. M. and BAZIN, M. J. (1983) A tubular bioreactor for photosynthetic production from carbon dioxide: design and performance, *J. Chem. Tech. Biotechnol.*, **33B**, 35.

PULZ, O. (1992) Cultivation techniques for microalgae in open and closed systems. In KRETSCHMER, P., PULZ, O. and GUDIN, C. (Eds) *Proc. 1st Europ. Workshop on Microalgal Biotechnology 'Algology'*, p. 61, Potsdam: Print Express K. Beyer.

RAMOS DE ORTEGA, A. and ROUX, J. C. (1986) Production of *Chlorella* biomass in different types of flat bioreactors in temperate zones, *Biomass*, **10**, 141.

RATCHFORD, I. A. J. and FALLOWFIELD, H. J. (1992) Performance of a flat plate, air-lift reactor for the growth of high biomass algal cultures, *J. Appl. Phycol.*, **4**, 1.

RICHMOND, A. (1986) Outdoor mass cultures of microalgae. In RICHMOND, A. (Ed.) *Handbook of Microalgal Mass Culture*, p. 285, Boca Raton, Fl.: CRC Press.

RICHMOND, A. (1990) Large scale microalgal culture and applications. In ROUND, F. E. and CHAPMAN, D. J. (Eds), *Progress in Phycological Research*, Vol. 7, p. 269, Bristol: Biopress.

RICHMOND, A. and GROBBELAAR, J. U. (1986) Factors affecting the output rate of *Spirulina platensis* with reference to mass cultivation, *Biomass*, **10**, 253.

RICHMOND, A. and VONSHAK, A. (1978) *Spirulina* culture in Israel, *Arch. Hydrobiol. Beih. Ergebn. Limnol.*, **11**, 274.

RICHMOND, A., BOUSSIBA, S., VONSHAK, A. and KOPEL, R. (1993) A new tubular reactor for mass production of microalgae outdoors, *J. Appl. Phycol.*, **5**, 327.

RICHMOND, A., LICHTENBERG, E., STAHL, B. and VONSHAK, A. (1990) Quantitative assessment of the major limitations on productivity of *Spirulina platensis* in open raceways, *J. Appl. Phycol.*, **2**, 195.

SOEDER, C. J., BOLZE, A. and PAYER, H. D. (1981) A rocking-tray for sterile mass synchronous cultivation of microalgae, *Br. Phycol. J.*, **16**, 1.

TORZILLO, G., GIOVANNETTI, L., BOCCI, F. and MATERASSI, R. (1984) Effect of oxygen concentration on the protein content of *Spirulina* biomass, *Biotechnol. Bioeng.*, **26**, 1134.

TORZILLO, G., PUSHPARAJ, B., BOCCI, F., BALLONI, W., MATERASSI, R. and FLORENZANO, G. (1986) Production of *Spirulina* biomass in closed photobioreactors, *Biomass*, **11**, 61.

TORZILLO, G., CARLOZZI, P., PUSHPARAJ, B., MONTAINI, E. and MATERASSI, R. (1993) A two-plane tubular photobioreactor for outdoor culture of *Spirulina*, *Biotechnol. Bioeng.*, **42**, 891.

TREDICI, M. R. and MATERASSI, R. (1992) From open ponds to vertical alveolar panels: the Italian experience in the development of reactors for the mass cultivation of phototrophic microorganisms, *J. Appl. Phycol.*, **4**, 221.

TREDICI, M. R., MANNELLI, D. and MATERASSI, R. (1988) Impianto perfezionato per la coltura in strato laminare sottile dei microrganismi fotosintetici, Italian Patent, ref. no CNR 9357.

TREDICI, M. R., CARLOZZI, P., CHINI ZITTELLI, G. and MATERASSI, R. (1991) A vertical alveolar panel (VAP) for outdoor mass cultivation of microalgae and cyanobacteria, *Bioresource. Technol.*, **38**, 153.

TREDICI, M. R., CHINI ZITTELLI, G. and BIAGIOLINI, S. (1992) Influence of turbulence and areal density on the productivity of *Spirulina platensis* grown outdoors in a vertical alveolar panel. In KRETSCHMER, P., PULZ, O. and GUDIN, C. (Eds) *Proc. 1st Europ. Workshop on Microalgal Biotechnology 'Algology'*, p. 58, Potsdam: Print Express K. Beyer.

TREDICI, M. R., CHINI ZITTELLI, G., BIAGIOLINI, S. and MATERASSI, R. (1993) Novel photobioreactors for the mass cultivation of *Spirulina* spp., *Bulletin de l'Institut Oceanographique de Monaco*, **12**, 89.

VONSHAK, A. and RICHMOND, A. (1988) Mass production of the blue-green alga *Spirulina*: an overview, *Biomass*, **15**, 233.

VONSHAK, A., ABELIOVICH, A., BOUSSIBA, S., ARAD, S. and RICHMOND, A. (1982) Production of *Spirulina* biomass: effects of environmental factors and population density, *Biomass*, **2**, 175.

WEISSMAN, J. C., RAYMOND, P. G. and BENEMANN, J. R. (1988) Photobioreactor design: mixing, carbon utilization, and oxygen accumulation, *Biotechnol. Bioeng.*, **31**, 336.

WYMAN, M. and FAY, P. (1987) Acclimation to the natural light climate. In FAY, P. and VAN BAALEN, C. (Eds) *The Cyanobacteria*, p. 347, Amsterdam: Elsevier Science.

8

Mass Culture of *Spirulina* Outdoors – The Earthrise Farms Experience

AMHA BELAY

Introduction

Even though *Spirulina* has a long history of human use (Dangeard, 1940; Cifferi, 1983), commercial production of *Spirulina* in man-made ponds was pioneered by Dainippon Ink & Chemicals Inc. (DIC) only as recently as 1978 in Bangkok, Thailand. Earthrise Farms was founded in 1981 by the then Proteus Corporation of the USA and was later incorporated with DIC of Japan in 1982. Commercial production of *Spirulina* at Earthrise Farms in Calipatria, California, was begun in the

Figure 8.1 Aerial view of outdoor ponds at Earthrise Farms, Calipatria, California.

summer of 1983. Earthrise Farms is the first *Spirulina* farm established in the USA. Today the farm is a subsidiary of DIC and supplies North and South America, Europe and Japan, as well as the Pacific Basin.

With a total pond area of 75 000 m^2 and an annual food-grade *Spirulina* production capacity in excess of 200 000 kg, this is the largest food-grade *Spirulina* plant in the world. There is a total of 15 production ponds, each having an area of 5 000 m^2 (Figure 8.1). In addition, the facility has about 40 000 m^2 of pond area for the production of feed-grade *Spirulina*. About 20 000 kg of feed-grade *Spirulina* is produced yearly. The farm also has numerous experimental ponds ranging from 2 m^2 to 1 000 m^2 in area.

The published information on large-scale production of *Spirulina* is rather scanty, and long-term data from large-scale commercial facilities are virtually absent. Either unavailability or secrecy of such data has barred producers from publishing the results of their experience in mass culture of *Spirulina* outdoors. The objective of this chapter will be to outline the experiences of Earthrise Farms in the mass culture of *Spirulina*, highlight some of the major problems encountered and discuss the present status and future prospects of commercial production of *Spirulina*.

Early Attempts in *Spirulina* Production

Spirulina *Producers*

During the last decade, several companies have attempted commercial production of *Spirulina*. Table 8.1 summarizes the status of past and present producers of *Spirulina*. The information is not complete and some of the data are estimates. However, they provide a general picture of the state of *Spirulina* production worldwide. Only those companies producing food-grade *Spirulina* are included. With the exception of Sosa Texcoco, the production area and annual production given on the table are for food-grade *Spirulina*. Many attempts in commercial production have failed while some successful producers are still limited to domestic markets. A few companies are important in the international market. Of the latter, excluding Sosa Texcoco, which is a semi-natural production facility, Earthrise Farms constitutes about 60 per cent of the total area and produces about 45 per cent of the total annual production sold to the USA and international market. The two farms belonging to Dainippon Ink & Chemicals (Earthrise Farms and Siam Algae Company) together constitute about 75 per cent of the total area and about 70 per cent of the total production.

The realization of the potential therapeutic effects of *Spirulina* based on current research findings in China and elsewhere (Belay et al., 1993) has attracted the attention of several Chinese producers in the last few years. Over the last couple of years alone, several facilities have been newly established in the People's Republic of China, while some others are under construction. The first commercial production of *Spirulina* from natural lakes was also started in Myanmar in 1988 (Thein, 1993). The increased awareness about the benefits of *Spirulina* for human and animal use will no doubt result in a steady increase in the establishment of new facilities all over the world.

Table 8.1 Current (1993) and past producers of *Spirulina* in the world

	Location	Area (m^2)	Annual Production (kg)
International Market			
Sosa Texcoco	Mexico	430 000	300 000[a]
Siam Algae Company	Thailand	20 000	75 000
Cyanotech	USA	33 600	110 000
Earthrise Farms	USA	75 000	143 000
Domestic Market			
Solarium	Chile	760	500
Cyanotech Bioproducts Ltd	India		10 000–12 000
Murugappa Chettir Research Center	India	5 000	
New Ambadi Estates Private Ltd	India	4 000	6 500
Sun Farms	Japan	13 500	30 000
Shenzhen Blue-Algae Biotechnology Corp.	P.R. China	12 000	10 000
Shenzhen Alginate Biological Company	P.R. China	12 000	33 000
Sanya Hainang Biological Develop. Co.	P.R. China	10 000	110 000
Chenghai Spirulina Co.	P.R. China		44 000
Wuhan Lanbao Microbial Algae Technology Joint Co.	P.R. China	10 000	110 000
Hainan Haiwen Pharmaceutical Research Center	P.R. China		110,000
Hainan Ocean Research Institute	P.R. China	2 000	13 200
Tian Microalgae Co.	P.R. China		55 000
Nanhua Bio-tech Development Labs	P.R. China	300	1 100
Provimin (harvested from lakes)	Myanmar		25 000–30 000
IMADE S. L.	Spain	5 000	
Nan Pao	Taiwan	66 000	200 000
Neotech Food Co.	Thailand	14 000	24 000
Failed Projects			
Proteal	French Indies	12 000	
Ein-Yahav	Israel	5 500	10 000
Koor Foods	Israel	12 000	17 000
Nippon Spirulina	Japan	13 500	60 000
Photo Bioreactor Ltd	Spain		
Blue Continent	Taiwan	30 000	60 000
Far East Microalgae	Taiwan	33 000	
Tung Hai	Taiwan	30 000	60 000
Cal Alga[b]	USA	30 000	
Spirutec[c]	USA		

[a] includes feed grade.
[b], [c] never fully established.

The limited number of *Spirulina* producers and the several abandoned algal projects point to the fact that large-scale commercial production of *Spirulina* has problems. As Becker (1994) has aptly put it '... the successful growth of algae is "high technology"; it is more or less an art and a daily tightrope act with the aim of keeping the necessary prerequisites and the various unpredictable events involved in algal mass cultivation in a sort of balance'.

Production of *Spirulina* at Earthrise Farms

Environmental Conditions For Spirulina *Cultivation*

Location of the Farm

Earthrise Farms is located in the Imperial Valley of California at approximately 33° N latitude and at an altitude of about 60 m below sea-level. It is situated in a remote and extensive agricultural area which derives its irrigation water from the Colorado River. The farm is located far from cities, highways and airports.

Temperature

The optimum temperature for the growth of *Spirulina* is around 35–38 °C, while the minimum temperature required for some growth is about 15 °C. There is virtually no facility that is located in an area where the optimal temperature is experienced throughout the year. Thus, production in outdoor ponds is for the most part limited by temperature. Production facilities located at lower latitudes like Thailand and Mexico can produce for about 10–12 months, while those at higher latitudes produce for about 6–7 months. The maximum, minimum and mean air temperatures recorded at the Calipatria Weather Station during 1984–1991 are given in Figure 8.2. Because of the low temperatures during the months of November to March, *Spirulina* production can proceed for only the 7 months from April to October. Even during the peak production period in summer there is a pronounced diurnal variation in temperature of about 20 °C. The effect of this diurnal variation on overall productivity has not yet been studied. While it can affect early morning photosynthetic productivity (Richmond et al., 1990), it may result in lower night

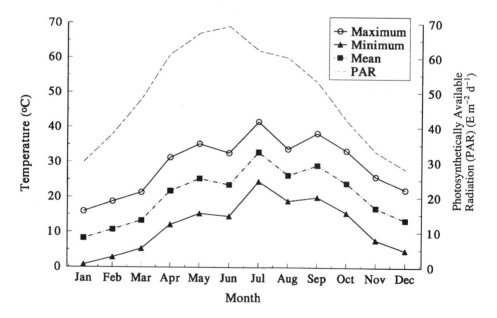

Figure 8.2 Seasonal distribution of temperature and irradiance at Calipatria, California (1984–1991).

time biomass loss than would be expected if the day temperature were to extend through the night (Torzillo et al., 1991; Vonshak et al., 1982). Another problem of low winter temperatures is the difficulty in maintaining healthy cultures in open outdoor ponds for use as seed culture for the following growth season. Under the cold winter temperatures, the culture often deteriorates fast and may eventually crash. For this reason, culture scale-up has to be done in greenhouse ponds of limited volume or in laboratory cultures, making the culture scale-up process slow and expensive. Since 1990, Earthrise Farms has successfully overwintered *Spirulina* culture in open outdoor ponds, making the culture expansion process faster and more economical.

Irradiance

The mean daily solar radiation (400–700 nm) falling on a horizontal surface of the Imperial Valley (El Centro Station) is given in Figure 8.2. The maximum irradiance is experienced in the months of May and June ($67-69$ E m^{-2} d^{-1}) whereas the minimum is recorded in December (28 E m^{-2} d^{-1}). Despite this, there is sufficient irradiation even in winter to allow reasonable production. Thus, production in winter is limited mainly by temperature ($8-15\,°C$).

Precipitation and evaporation

The average annual rainfall is less than 80 mm. In the period from November to March, the 'rainy' season is, on the average, 16 h with rain – a little more than 3 h per month – making this area a decidedly desert region. The average daily evaporation is about 6 mm. There is thus a significant amount of water deficit which, in the case of the pond culture, has to be replaced daily with fresh water.

The environmental constraints under which the farm is operating are thus characterized by (a) low winter temperatures limiting productivity to only 7 months per year, (b) the pronounced diurnal variation in temperature during the growth season, possibly resulting in decreased growth rates in the early morning hours, and (c) the high rate of evaporation resulting in high volumes of water replenishment with the attendant precipitation and accumulation of $CaCO_3$.

Mass Cultivation of Spirulina

Outline of the production system

The *Spirulina* production system can be broadly divided into four phases: culturing, harvesting, drying, and packaging. Each of these processes is important in the production of high-quality *Spirulina*. A schematic diagram of the production system at Earthrise Farms is presented in Figure 8.3.

The factors affecting the production of *Spirulina* in outdoor ponds have been the subject of extensive study and review (Richmond et al., 1990; Torzillo et al., 1991; Vonshak et al., 1982; Vonshak, 1987a; Goldman, 1978). A detailed discussion of the growth conditions at Earthrise Farms is thus beyond the scope of this chapter. In the following, a brief description is given of the processes and some of the problems associated with each phase of the production system.

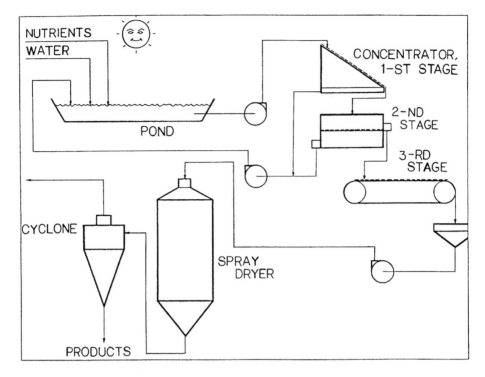

Figure 8.3 A schematic diagram of the *Spirulina* production system.

Design and mixing of ponds

A detailed description of the basic design of the ponds has been given by Shimamatsu (1987). The culture is grown in PVC-lined raceways of 5000 m² area. Mixing is introduced by means of single or double paddle wheels. The depth of the ponds depends on the season, desired algal density and, to a certain extent, the desired biochemical composition, particularly that of pigment concentration. Therefore, depth can vary between 15 and 30 cm. It is generally known, from this and other *Spirulina* production facilities, that paddle wheel mixing does not provide sufficient turbulence to give the algae the optimum residence time in the photic zone. Flow velocities of the order of 60 cm s^{-1} have been recommended. However, in our experience, increasing the flow velocity above 30 cm s^{-1} invariably results in fragmentation and increased coiling of trichomes which gradually lead to decreased efficiency of harvest and reduced yield. Increasing flow velocity above 30 cm s^{-1} also results in increased expenditure for pond maintenance and energy (Oswald, 1988). The optimum flow velocity allowing high growth rates depends on the operational depth of the pond, the algal density maintained and the physiological state of the algae.

The culture medium

The culture medium is based on the original medium of Zarouk (1966) and is typically composed of sodium carbonate, a source of nitrogen, phosphorous, iron and trace metals. After harvesting, the culture water is recycled continuously into the

136

ponds throughout the growth season. Only food-grade chemicals are used. The choice of chemicals and concentrations that are used depends on cost and the compatibility of the chemicals with the raw water. Cost of nutrients accounts for about 15–25 per cent of the total production cost. The farm has done extensive research trying to develop a medium concentration and supply regime that provides optimum growth rates at minimum cost. A comparison of the relative cost of production over a 10-year period (see Figure 8.10) shows a progressive decline in unit cost of production. This reduction in cost of production is partly due to improvements in the composition and quantity of nutrients used. The latter is based on a laboratory and outdoor study of the requirements of the algae for specific nutrients.

One major problem encountered during outdoor culturing of *Spirulina* is the high calcium content of the raw water. This results in the precipitation of $CaCO_3$ during the initialization of the culture and during daily additions of freshwater to make up for evaporative losses. This phenomenon results in the reduction of alkalinity and to a certain extent loss of iron and phosphorus from the system. Pretreatment of the raw water during the initialization process and continuous removal of detritus throughout the growth season are among the steps taken to alleviate these problems.

Strain selection

The major determinants in the selection of strains for commercial production are growth rate, biochemical composition and resistance to mechanical and physiological stress. A wide variety of species and strains of *Spirulina* have been screened by several people in various countries. Notable among these are those done in Israel (Vonshak, 1987b) and in Japan (DIC, unpublished data). During the course of its operation, Earthrise Farms, in cooperation with its parent company DIC, has made extensive collections of *Spirulina platensis* and *Spirulina maxima* from various localities around the world. A large number of strains have been screened for desirable traits as mentioned above and for their suitability to be grown in mass culture under the prevailing conditions of the farm. While the selection of suitable strains and scaling-up of the culture is relatively easy, the harvesting and recycling processes result in cross-contamination thus precluding the maintenance of pure strains throughout the growth season. This is particularly so if several strains are grown and harvested during the same period. The problem can be minimized through the use of the same strain throughout the growth season or some defined period thereof. It is often difficult to tell if one is using the same strain, since the morphology is similar. However, certain biochemical markers can be used to give a rough idea of the composition of the population. For example, total carotenoid content, phycocyanin content and susceptibility to certain stress factors like photoinhibition or mechanical damage can be monitored and compared against known levels for each strain.

Scale-up of culture

One of the most difficult tasks in outdoor mass culture is the scaling-up process. This is the stage in the process where contamination by other algae and bacteria poses the greatest problem because of the initial dilute nature of the *Spirulina* inoculum. According to Richmond (1990) and unpublished results from Earthrise Farms, there seems to be a direct relationship between the density of *Spirulina* in the culture and the density of contaminants. Contamination by green algae is high when the initial density of the inoculum is low. Conversely, the amount of contamination decreases as the *Spirulina* culture builds up in density. It is speculated that extracellular products of *Spirulina* may have some allelopathic properties. Threshold concentration of these substances is probably reached at high cell densities. Light limitation of green algae by the positively buoyant trichomes of *Spirulina* may also account for the observed phenomenon. Through careful manipulation of the nutrient concentration and useful natural predators, it has been possible to maintain a unialgal culture even during the initial period of inoculation.

Earthrise Farms utilizes three modes of expansion. In the first, culture is scaled up from strains maintained in test tubes. The scale-up follows a roughly $1:5$ dilution ratio through successive volumes up to the 1000 m^3 culture in the production ponds. In the second mode of operation, the scale-up operation utilizes culture from the previous growth season maintained in greenhouses. Since 1992, Earthrise Farms has successfully developed a method of overwintering culture in open production ponds. Properly maintained, this third mode of operation provides an inoculum as viable as the other two kinds of operation. In this last mode of operation, culture expansion to the entire volume of the 15 production ponds ($15\,000 \text{ m}^3$) can be done from just a couple of ponds in less than a month.

Culture Maintenance

Proper culture maintenance calls for a routine monitoring of various physical, chemical and biological parameters. In the following a brief description of the culture maintenance procedures used at Earthrise Farms is presented.

pH control

The pH of the medium is one of the most important factors in culturing *Spirulina*. Maintaining a pH of over 9.5 is mandatory in *Spirulina* cultures in order to avoid contamination by other algae. pH adjustment is made by supplying CO_2 gas to the medium. As in other large-scale mass-culture facilities, the CO_2 supply or inlet system employed is a compromise between efficient gas transfer, capital cost, and the operation cost of the system. During the course of its operation, Earthrise Farms has employed different CO_2 supply systems, ranging from plastic dome exchangers to air stones and perforated PVC pipes (Becker, 1994; Bonnin, 1992). All these three types of systems are economical to use. However, the highest efficiency of transfer achieved to date at the farm using any of these systems is only about 60 per cent. The main areas of loss are exchange with the atmosphere, precipitation as $CaCO_3$ and loss to the part of the medium that is not recycled back to the ponds. The relative contribution of the above factors to the overall loss of CO_2 from the system has not yet been evaluated. CO_2 injection is made by manually controlling the valve pressure. Currently, the farm is studying alternative methods of CO_2 injection and there are plans to automate the system. It is rare that the pH of the medium falls below about 9.0. However, upward shifts in pH in excess of pH 10.5 are quite common due to human error. When this occurs, it is accompanied by precipitation of $CaCO_3$, which is sometimes followed by flocculation and sedimentation of algae.

138

Excessive deposition of such detritus has undesirable effects on the culture. For example, the resuspension of the detritus contributes to light attenuation. In extreme cases, light absorption from such non-algal components can be as high as 25 per cent of the incident light. Moreover, the decomposition of the organic matter results in the increase of bacterial population; in particular, ciliates seem to thrive well under these conditions. Under such circumstances, the culture has to be replaced completely to comply with quality guidelines.

Nutrient concentration

In continuous cultures, the actual concentration of nutrients in the medium is determined by the biomass concentration and the output rate, as well as by other losses in the system like precipitation or flow out. Moreover, the relative composition of the various nutrient elements may be altered from that of the initial medium by the changing biological, chemical and physical environments in the pond culture during the course of growth. Routine chemical analysis should be done for most of the major nutrients like nitrogen, phosphorus, potassium and magnesium, to prevent depletion. The fate of iron and trace metals in the alkaline medium of *Spirulina* culture is poorly understood. Because of the high content of Ca^{++} ions in the raw water of the medium, precipitation of $Ca_3(PO_4)_2$ and probably $FePO_4$ is a possibility. This is inferred from a constant loss of P and Fe from the medium that is unaccountable by growth and other known losses. It is estimated that close to 20 per cent of the phosphorus and iron is lost to such precipitation during the initialization of the culture. Apart from loss of nutrients, such a precipitate and its associated detritus can affect light penetration. At times, particularly toward the end of the growth season, light attenuation due to non-algal sources can be as high as 25 per cent. Partial recycling of these nutrients is possible, though the high pH, high alkalinity and high oxygen concentration may inhibit such a transfer. In our earthen ponds used for the production of feed-grade *Spirulina*, such a transfer of nutrients takes place at the sediment–water interface in the somewhat anoxic layers of these incompletely mixed ponds. Trace elements are difficult to analyze chemically. Consequently, the farm employs a bioassay technique to determine limitation by all or some components of the trace element mixture.

As mentioned earlier, nutrient cost amounts to about 15–25 per cent of the total production cost. Detailed knowledge of the nutrient uptake kinetics of *Spirulina* in large-scale open pond systems and the fate of certain nutrients in the high-pH and high-oxygen environment of the medium, is lacking. Such information, when available, will no doubt help minimize nutrient cost and/or increase productivity.

Control of depth and the underwater light environment

The maintenance of optimum depth is an important feature of large-scale systems through its effect on the mixing of ponds and on light penetration. The amount of light that impinges on the cells depends on the amount of surface radiation, the depth of the culture, the degree of turbulence and the population density (Vonshak et al., 1982; Richmond and Vonshak, 1978).

Light is an important limiting factor in dense algal cultures of the type normally encountered in open outdoor ponds. For good harvest efficiencies and hence high areal output of biomass, we find, as have other workers (Richmond et al., 1980), that the optimal biomass has to be in the range of 400–500 mg dry weight l^{-1}. At such densities, the residence time of the algae in the photic zone is reduced by about 75 per cent. Thus, the culture in outdoor open ponds is indeed light limited. At densities lower than 250 mg dry weight l^{-1}, harvest efficiency is reduced, as well as growth of the culture, most likely due to photoinhibition. This latter phenomenon has been demonstrated in various field and laboratory studies (Belay, 1981; Belay and Fogg, 1978) and is the subject of several reviews (Neale, 1987) but the extent to which it affects outdoor *Spirulina* production has only recently been elucidated (Vonshak and Guy, 1988).

The challenge that confronts large-scale outdoor *Spirulina* producers is how to maximize the amount of light received by individual cells. Several methods can be employed though none of these fully produces the desired effect. Lowering the density of the biomass increases light penetration and results in increased growth rate. However, yield of biomass per unit harvest time decreases at low biomass densities, resulting in the need to process a large volume of culture thereby increasing process time. Also, at such low densities, photinhibition can affect production, at least in summer. Another method used to solve the problem is to reduce depth. This may work, provided that mixing by the paddle wheel is not affected by the shallow depth, which is the case in fixed paddle wheel designs. Increasing the mixing rate of the culture increases the growth rate as well as the optimal biomass concentration (Richmond and Grobbelaar, 1986). It is generally known, from this and other *Spirulina* production facilities, that paddle wheel mixing does not provide sufficient turbulence to give the algae the optimum residence time in the photic zone. However, as mentioned earlier, increasing the mixing rate of the paddle wheels beyond a certain limit has some undesirable consequences.

Maintaining a unialgal culture

The maintenance of unialgal cultures of *Spirulina* throughout the growing season is a challenge to any commercial production facility (Vonshak and Richmond, 1988; Vonshak et al., 1983). The economic production of *Spirulina* necessitates the continuous recycling of the nutrients after the biomass is removed. The constant recycling of the medium often results in excessive accumulation of organic matter. This not only results in contamination by other algae (Richmond et al., 1990; Vonshak, 1987a), but also in what appears to be the autoinhibition of growth of *Spirulina*. The build-up of organic matter is also manifested at times in excessive foaming, which is a result of decomposition and death of algae. Earthrise Farms has succeeded in minimizing this problem by improving the harvest facility, culture monitoring and culture control conditions. These improvements have substantially reduced the breakage and fragmentation of cells. The culture water can therefore be recycled continuously throughout the 7-month production season without undue accumulation of organic matter.

The major contaminant algae in *Spirulina* ponds, as reported by Vonshak and Richmond (1988), are *Chlorella* and a small species of *Spirulina*, *Spirulina minor*. These workers have estimated the overall annual loss of productivity due to contamination and subsequent discarding of culture to be in the order of 15–20 per cent. The major contaminant alga at Earthrise Farms is probably the unicellular green alga *Oocystis sp*. It is an alkalophile like *Spirulina*, and increasing the alkalinity or pH does not have much effect on it. Prior to 1991, contamination by this alga occurred during the colder months of the year toward the beginning and end of the harvest season when temperature conditions favor its growth. Contamination by this alga contributed to as much as 25 per cent of the biomass as measured by optical density difference between total absorbency of the culture and after the same water had been filtered through a screen with a pore diameter of 40 mm (Figure 8.4). In 1986, for example, 5 out of 10 ponds were wasted in just one month because of excessive contamination by green algae (Figure 8.4 inset). Since 1990, Earthrise Farms has been effectively utilizing a proprietary biological method to control green algal contamination.

In culture that is managed carefully, contamination by zooplankton is minimal. Under poor culture conditions, the major contaminants are ciliates. These become abundant in the colder months when environmental conditions and mechanical stress render *Spirulina* susceptible to breakage and decomposition.

Insect control

Aquatic insects are unavoidable in open ponds. The groups of major importance are the Ephydridae (brine-flies), Corixidae (waterboatmen) and Chironomidae (midges). In terms of their contribution to insect fragments in the product, the ephydrids are the most abundant in our ponds and have been reported to be in other places too (Venkataraman and Kanya, 1981). The larvae of these flies have a large number of minute sclerotized spinules and crochets which further break down into still smaller pieces during the harvest process. The corixids are often large enough to be retained by pond netting. In addition to aquatic insects that may breed *in situ*, contamination by terrestrial insects or their parts also occurs, though not significantly. These and the aquatic insects are controlled by netting *in situ* and during preharvest screening.

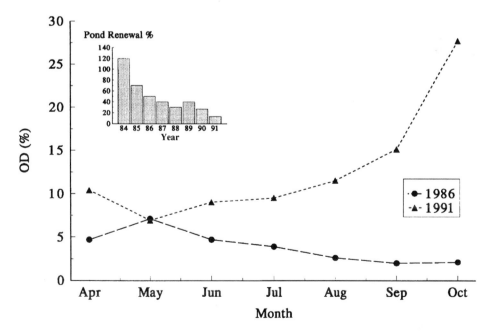

Figure 8.4 Green algal contamination as measured by optical density (OD). Inset: Number of ponds renewed as a percentage of the initial total number of ponds. See text for details.

141

Spirulina platensis (Arthrospira)

Rodent control

As will be mentioned later, USFDA defect action level (DAL) for rodent hair is 0.5 per 50 g of dry powder. It is therefore imperative to control the entry of rodents into the vicinity. The area surrounding the ponds is disked to prevent accidental entry of rodents. Rodent traps are also set around the ponds. Despite this, rodent hairs are transported by wind from the surrounding agricultural areas, especially during the burning of the fields after harvesting. Further removal of light filth is effected by netting *in situ* and during preharvest operations.

Harvesting the Biomass

The harvest system

The harvest system in *Spirulina* production plants usually involves a few stages of filtration. As shown in Figure 8.3, Earthrise Farms utilizes three stages of filtration to harvest the biomass. The final slurry is normally about 15–20 per cent solids.

Harvest efficiency

The efficiency of harvest (harvested biomass/processed biomass) depends on the trichome size and the mesh size of the filters used at each stage. The smaller the mesh size, the higher the efficiency. However, flow rates are invariably lower at the higher efficiencies associated with smaller-mesh filters. Increasing the force of the water flow often results in breakage of cells and hence loss in efficiency, and the return water has undesirable consequences in the pond culture. Nevertheless, a certain optimal flow rate must be achieved in order to remove most of the daily production. If this is not achieved, the build-up of biomass in the ponds results in reduced light penetration and reduction of growth rates and productivity. These circumstances can eventually lead to the unavoidable death and decomposition of the algae. Downstream filtering systems may also be affected if the flow rate in the upstream system is higher than a given value. The final screen configuration in the various stages is therefore a tradeoff between efficiency and flow rate and must be tuned up depending on trichome size and the total amount of biomass to be removed. Thus, culture control depends very much on harvest efficiency and *vice versa*.

Harvest-based versus in situ growth rates

To date, there is no harvest system that allows the complete removal of the incoming biomass at flow rates that are adequate enough to remove the desirable amount of biomass in large-scale commercial plant. An unavoidable problem during the harvest process is the mechanical damage to the culture induced by transporting and harvesting equipment. This damage manifests itself in the fragmentation of the trichomes which will eventually pass through the filtration equipment. In addition to fragmentation, there is also a progressive increase in the coiling of the trichomes and consequently a progressive decrease in the size of the trichomes during the harvest season. Growth rates calculated from the harvested biomass are about 30 per cent lower than those calculated from *in situ* changes in biomass (Figure 8.5). Part of this loss is due to death and sedimentation of the damaged biomass in the recycled water.

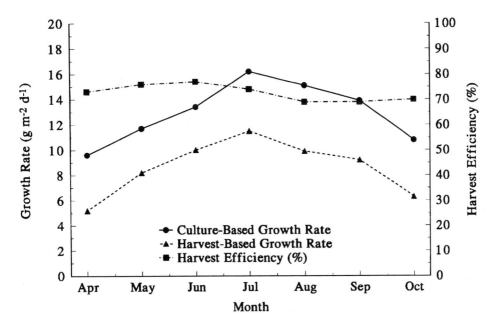

Figure 8.5 The relationship between *in situ* growth rate, harvest-based growth rate and harvest efficiency.

This coiling and reduction in trichome size results in a reduction of harvest efficiency toward the end of the production season (Figure 8.5). The efficiency of harvest calculated as the ratio between the total biomass harvested and the total biomass processed (dry weight) also shows about a 30 per cent reduction.

Harvest-induced stress and recovery in Spirulina

Dissolved organic matter and autoinhibition   The economic production of *Spirulina* necessitates the continuous recycling of the nutrients after biomass removal. The constant recycling of the medium often results in excessive accumulation of organic matter because of the fragmentation of the trichomes by the transport and harvesting systems. This not only results in contamination by other algae and bacteria but in what appears to be the autoinhibition of growth. Figure 8.6 shows the result of a simple bioassay of growth of *Spirulina*, conducted with filtered water taken from a pond that had to be wasted because of excessive contamination by green algae. The bioassay involving treatment of the filtered pond water with charcoal to remove organic matter showed a much higher growth rate of *Spirulina* than the untreated control sample, suggesting that growth of *Spirulina* was possibly inhibited by the accumulation of certain organic substances. As discussed earlier, improvements in culturing and harvesting conditions have resulted in reduced coiling and fragmentation of cells. Indeed, charcoal bioassay studies, conducted at the end of the 1991 production season, did not show any difference in growth rate of *Spirulina* compared with untreated controls (Figure 8.6).

Fragmentation and increased coiling of trichomes   A very common phenomenon observed at Earthrise Farms is the gradual change in the size and morphology of

143

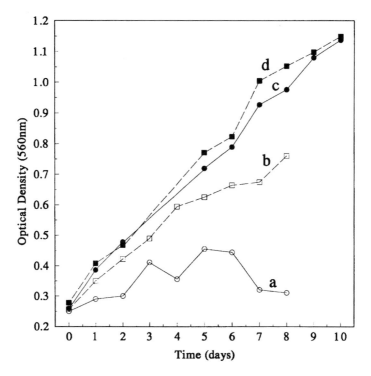

Figure 8.6 Growth of *Spirulina* in pond culture filtrate before (O) and after (□) treatment of the medium with activated charcoal. Open circles and boxes for 1989; closed circles and boxes for 1991.

trichomes with time or increasing frequency of harvest. There is a tendency for increased coiling as the growth season progresses (Figure 8.7). It seems to be initiated during the warmer months of the season but thereafter progresses even into the colder months of the production season. This coiling and reduction in trichome size results in a reduction of harvest efficiency toward the end of the production season (Figure 8.7). In cultures that are left to overwinter in an open pond, there is a gradual reversal to the long and loose coil type, whereas cultures transferred to heated greenhouses retain the tight-coiled structure. Such changes in the morphology of trichomes of *Spirulina* have been reported by several workers (Bai, 1980; Bai and Seshadri, 1980; Lewin, 1980). According to Bai and Seshadri (1980), high light intensity and high nutrient concentration caused the transformations from the loose-coiled variant (S-type) to the tighter coiled one (C-type), and the latter showed increased coiling to the very tight variant (H-type) at high light intensity and low nutrient concentrations. In our case, neither light nor nutrients varied significantly during the period when the transformation was noticeable. It thus seems that high temperature and/or mechanical stress may be the factors involved. The low light condition, low temperature and absence of harvest-related mechanical stress may be responsible for the reversal to the loose trichome condition in winter.

In an effort to understand this phenomenon better, a series of experiments were conducted to study the effect of mechanical damage on growth and photosynthesis. The results of these studies have been reported elsewhere (Belay, 1993). As shown in Figures 8.8 and 8.9, both photosynthesis and growth are severely affected

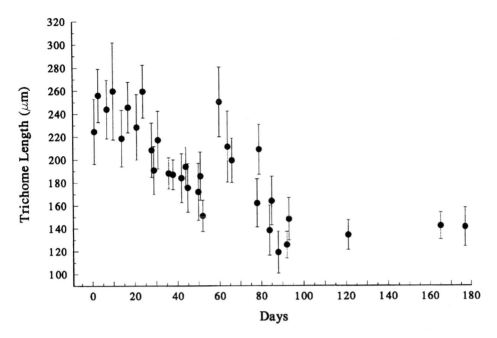

Figure 8.7 Temporal variation in trichome length.

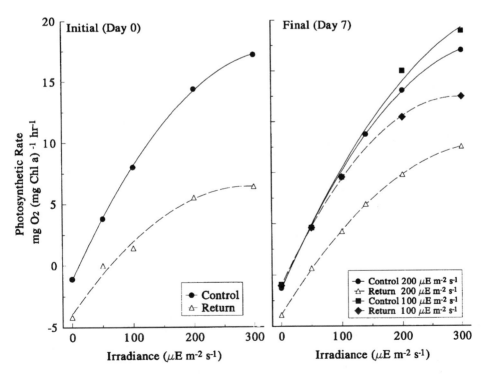

Figure 8.8 Photosynthesis–irradiance relationship of *Spirulina* immediately after mechanical damage (Day 0) and after 7 days of recovery (Day 7).

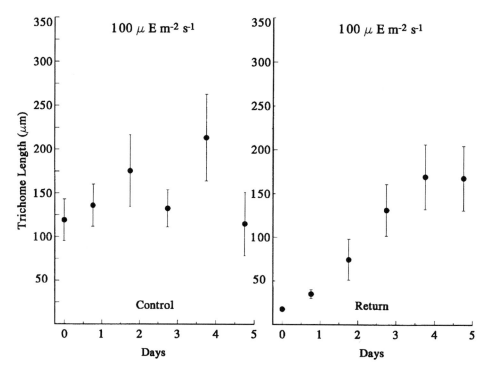

Figure 8.9 Trichome length as a function of time before (Control) and after (Return) mechanical damage.

immediately after the damage. However a certain degree of recovery was observed after incubation for a few days. Although recovery of the trichomes *in situ* can be envisaged on the basis of the above observations, the extent of this recovery has not as yet been evaluated.

Drying and Packaging

Significance in quality control

Proper and quick drying is an essential feature of high-quality *Spirulina* production. Various types of drying systems are used in the industry for drying *Spirulina*. For economic reasons, the dryer of choice in large-scale *Spirulina* production facilities is the spray dryer. Freeze drying would give better overall product quality, but the cost is rather prohibitive. Earthrise Farms utilizes spray drying. *Spirulina* droplets are sprayed into the drying chamber just long enough to flash evaporate the water. The powder is exposed to 60 °C heat for a few seconds as it falls to the bottom. No preservatives, additives or stabilizers are used in drying, and it is never irradiated. This quick spray drying process guarantees preservation of heat sensitive nutrients, pigments and enzymes. In line with good manufacturing practices, the product is not handled by human hands during harvesting, drying and packaging.

Proper packaging is another important feature of high-quality *Spirulina* production. The powder coming out of the dryer is immediately vacuumed away to a collection hopper in the packaging room and is sealed under vacuum in drums with special gas-barrier bags in order to minimize oxidation of certain vital pigments like carotenoids. The sealed product is then placed in drums for shipment. In these drums, the product can stay up to 4 years with little change in biochemical composition or nutritional properties. The choice of packaging material is based upon studies involving temporal variations in pigment loss under different storing conditions. The results of one such study are given in Table 8.2. Under a gas-barrier bag there appears to be no further loss after the initial oxygen concentration in the bag is exhausted.

Moisture content

Improper drying often results in high moisture content in the product. A moisture content in excess of 8 per cent will result in the growth of molds and bacteria in the product. Quick and efficient drying is effected by using a spray dryer. Since the dried powder has a high sorption characteristic, the product is immediately packaged in a dry environment. The maintenance of an optimum temperature in the dryer is an important aspect of the drying process. Overdrying often results in the loss of some essential components like vitamins, chlorophyll and carotenoids. For this reason the dryer at Earthrise Farms is microprocessor controlled, so that the drying temperature is constantly maintained and controlled.

Bulk Density

The most important physical characteristic of *Spirulina* powder is the bulk density. This is because it affects the tabletability of the product. Generally there are two types of tableting processes, one requiring a high bulk density product and the other requiring a low bulk density. It is therefore very difficult for a single production facility to satisfy the requirements of varied customers. The bulk density of the product is affected by particle size distribution, type of agglomeration, particle

Table 8.2 Pigment degradation (percentage of initial) under different temperature and package conditions

Time (months)	Phycocyanin 25 °C	50 °C	Carotenoids 25 °C	50 °C	Chlorophyll 25 °C	50 °C
Polythylene bag						
0.0	100	100	100	100	100	100
0.5		87		59		98
1.0	108	93	96	44	105	96
2.0	97	75	88	29	96	87
5.0	89	57	50	10	95	74
Gas-barrier bag						
0.0	100	100	100	100	100	100
0.5		96		99		101
1.0	96	109	102	97	100	100
2.0	97	83	102	98	100	98
5.0	95	88	98	97	100	97

porosity, and to a certain extent the moisture content. Particle size distribution is affected by the initial size of the trichomes as they are fed to the dryer and the pore diameter of the atomizer. The final quality of the product with respect to bulk density is therefore dependent on culturing, harvesting and drying conditions. To a certain extent, all these factors are harnessed in order to obtain a product that meets the requirements of various customers.

Quality Assurance

Routine analysis of the chemical, physical, biochemical and microbiological characteristics of the product is of paramount importance in food-grade *Spirulina* production. A well-equipped laboratory and trained personnel are essential since it is advisable and economical to do the analysis on site. Outside laboratory analysis is costly and often unreliable. Earthrise Farms does most of the chemical and microbiological analysis on site. Occasionally, samples are sent to independent laboratories for confirmation and standardization. This is necessary in order to ascertain that the final product meets quality criteria for domestic as well as international markets. Only after each production lot has passed all tests is it certified and ready to ship.

Chemical Composition

One aspect of high-quality *Spirulina* production is that the product must have a consistent chemical and physical property. Table 8.3 provides a typical chemical composition of *Spirulina* powder produced at Earthrise Farms, together with some physical characteristics. The analysis was made by Japan Food Association Laboratories on mixed samples of products representing the entire production of the 1988 and 1992 growth season. Seasonal and annual variations in product quality are observed. However, when one considers the open nature of the production system, the consistency of biochemical composition is remarkable (Belay and Ota, 1994). This may show that, while there is no doubt that product quality and consistency will be much better in algae grown in closed bioreactors, the variation in product quality of algae grown in well-maintained open ponds is not as high as one would expect. Good culture and product management conditions are prerequisites for product quality and uniformity of biochemical composition.

Microbiological Standards

The cultivation of microalgae in large open ponds using surface water will undoubtedly invite contamination of the harvested product by other microorganisms. Subsequent handling of the product during harvest, drying and packaging could also result in microbial contamination. The final microbial load of the product will therefore depend on how carefully the culture and product are handled at the various stages. Direct examination of the microbial flora in each lot of dried product is therefore essential. Standard Plate Counts (SPC) and confirmed coliform counts are used in the food industry to monitor and inspect malhandling of food products during

Table 8.3 Chemical and physical properties of Earthrise Farms *Spirulina*

General Composition	(%)	Amino Acids	(g kg^{-1})
Moisture	3–7	Alanine	47
Protein	55–70	Arginine	43
Fat (Lipids)	6–8	Aspartic acid	61
Carbohydrate	15–25	Cystine	6
Minerals (Ash)	7–13	Glutamic acid	91
Fiber	8–10	Glycine	32
		Histidine	10
Vitamins	(mg kg^{-1})	Isoleucine	35
		Leucine	54
Provitamin A	2 330 000 IU kg^{-1} [a]	Lysine	29
(β-carotene)	1 400	Methionine	14
Vitamin E[b]	100	Phenylalanine	28
Thiamin B-1	35	Proline	27
Riboflavin B-2	40	Serine	32
Niacin B-3	140	Threonine	32
Vitamin B-6	8	Tryptophan	9
Vitamin B-12	3.2	Tyrosine	30
Inositol	640	Valine	40
Folic acid	0.1		
Biotin	0.05	*Essential Fatty*	(g kg^{-1})
Pantothenic acid	1.0	*Acids*	
Vitamin K-1	22		
		Linoleic acid	8
Minerals	(mg kg^{-1})	γ-linolenic acid	10
Calcium	7 000	*Pigments*	(g kg^{-1})
Chromium	2.8		
Copper	12	Carotenoids	3.7
Iron	1 000	Chlorophyll	10
Magnesium	4 000	Phycocyanin	140
Manganese	50		
Phosphorus	8 000	*Enzymes*	(Units kg^{-1})
Potassium	14 000		
Sodium	9 000	Superoxide dismutase	1 500 000
Zinc	30		

Physical Properties

Appearance	fine powder
Color	dark blue-green
Odor and taste	mild like seaweed
Bulk Density	0.35–0.55 kg l^{-1}
Particle size	64 mesh through

[a] not mg/kg;   [b] a-tocopherol equiv.

processing (Eisenberg, 1985; Jay, 1992). Ten years of tests at Earthrise Farms show that there are almost no positive confirmed coliform found, indicating the generally good sanitary conditions of growth, harvest, and drying and packaging conditions. Studies on products from Chad, Algeria and Mexico have indicated high coliform counts (Jacquet, 1976). However, analysis of hundreds of *Spirulina* samples from modern commercial farms in Thailand, Japan, Taiwan and Mexico show that coliforms are rarely present (Jassby, 1988a). The use of these organisms as indicator species can be misleading since other forms of pathogenic organisms may still be present. It is therefore generally advisable to follow good manufacturing practices in order to avoid contamination at all stages of the production system.

The final microbiological load should conform to the standards set by different countries where the product is to be marketed. Table 8.4 shows a comparison of the microbiological standards of different countries with that of Earthrise Farms internal quality standard. The average values of quality control data obtained for the years 1989–1993 are much lower than those of even the farm's own strict standard. A note should be given here concerning standard plate count. Standard plate counts of *Spirulina* are known to decrease an order of magnitude after only a couple of months of dry storage (Jacquet, 1976). Hence even the strict SPC guideline of France can be met.

Extraneous Material

According to AOAC (Association of Official Analytical Chemists) (1990a), extraneous material is the name given to 'any foreign matter in product associated with objectionable conditions or practices in production, storage, or distribution'. If the extraneous matter is contributed by insects, rodents, birds or other animal contamination, it is referred to as *filth*. The major components of extraneous matter in food products are insect fragments, rodent hair, and feather fragments.

Table 8.4 Microbiological and related quality standards in relation to the quality of Earthrise Farms *Spirulina*

	France[a]	Sweden[b]	Japan[c]	USA[d]	EF[e]	EF (89–93)[f]
Moisture (%)			<7.0		<7.0	4.3 ± 0.4
SPC ($\times 10^6$)/g	<0.1	<10.0	<0.05		<1.0	2.17 ± 1.36
Mold (#/g)		<1000	<100		<100	<11
Yeast (#/g)					<40	Neg
Coliforms (MPN/g)	<10	<100	Neg		Neg	Neg
Salmonella	Neg	Neg	Neg		Neg	Neg
Staphylococcus (#/g)	<100	<100			Neg	Neg
Insect Fragments (#/10g)				<30	<500	48 ± 32
Rodent Hair (#/150g)				<1.5	<1.5	0.1 ± 0.2

[a] Superior Public Hygiene of France
[b] Ministry of Health, Sweden
[c] Japan Health Food Association
[d] United States Food and Drug Administration
[e] Earthrise Farms Standard
[f] 1989–1993 data (average)

EF = Earthrise Farms
Neg = Negative
SPC = Standard Plate Count
MPN = Most Probable Number

In conventional foods, insect fragments often indicate improper manufacturing and/or storage conditions. This can hardly be interpreted as such in microalgal products, since aquatic insects are indigenous to outdoor microalgal ponds. As indicated above, brine flies (Ephydridae), waterboatmen (Corixidae) and midges (Chironomidae) are common in *Spirulina* ponds. The presence of fragments from indigenous insects in the dry powder should present no hazard to physical health. Indeed *Ephedra hians*, which is the dominant aquatic insect in our ponds, has been used as food by the Indians in the United States and Mexico (Aldrich, 1912; Wirth, 1971). The Paiute Indians of California and Nevada gave this food the name of 'Koo-chah-bee', while the Aztecs of Mexico called the larvae 'Puxi'.

A standardized analytical method exists for counting insect fragments (AOAC, 1990a, b). Earthrise Farms has collaborated in the development of this official method of analysis for insect fragments in *Spirulina* (Nakashima, 1989). While the sample preparation and counting techniques are very well standardized, a lot is to be desired for the identification process. The author has sent the same preparations to three different laboratories and found results varying by two orders of magnitude. Most of the literature on identification of insect fragments is based on agricultural or storage insects. In addition, it is often difficult to distinguish between insect parts and plant parts, resulting in an overestimation of 'unidentified insect' parts. Researchers at Earthrise Farms and the University of Texas at Austin are developing a method to quantify insect fragments using a myosin ELISA technique.

The relatively high average insect fragment counts in Earthrise Farms products (1989–1993) is due to the relatively large number of insect counts that were observed during the first four years of this period. Subsequent improvements in insect control have resulted in relatively lower numbers of insect fragments in the product. In 1993, for example, all food-grade products had insect fragment counts of less than 30, which is the USFDA requirement. Over 95 per cent of the insect parts recorded are usually of aquatic insect origin. If insect parts from indigenous sources were to continue to be considered unsanitary, then many *Spirulina* producers would have difficulty in selling their products in the USA. The cost of controlling aquatic insects would be prohibitive to many small-scale *Spirulina* producers.

The presence of rodent hair in microalgal products is considered to be an indicator of potential contamination. At Earthrise Farms the source of rodent hair is probably the deer mouse *Peromyscus maniculatus*. There are several ways of entry of rodent hair into the product. Though never observed at Earthrise Farms, accidental entry of rodents into production ponds can occur if proper rodent control measures are not implemented. Another source of rodent hair is transport by wind from the surrounding areas, especially during the burning of fields after harvest. Data from Earthrise Farms show a direct correlation between wind speed and rodent-hair levels in harvested algae. We have also observed that there is a significant correlation between control measures like trapping and disking and the number of rodent hairs in the product.

When proper rodent control measures are taken, very rarely would rodent hair be observed in the dry product. Currently, Earthrise Farms has no problem in preventing contamination of its products by rodent hair. Almost all the production lots from food-grade production meet the current USFDA guideline of 1.5 rodent hairs per 150 g of *Spirulina* (Table 8.4). Feathers, plant fragments and any other extraneous material are strained during pond netting and preharvest screening. The product is therefore relatively clean of extraneous material.

Spirulina platensis (Arthrospira)

Heavy-Metal Standards

Lead, mercury, cadmium and arsenic are the major contaminants of microalgal products since they are components of industrial pollution and occur in trace amounts in certain pesticides and agricultural fertilizers. *Spirulina* is also known to concentrate heavy metals in excess of the concentrations found in the medium (Lacquerbe et al., 1970). The production of high-quality *Spirulina* therefore requires the use of high-grade nutrients, which are costly. It is also necessary to conduct routine or periodic analysis of heavy metals in the product. This is particularly important in situations where food-grade *Spirulina* is to be produced from earthen ponds or natural lakes. The soil in certain regions may have a high content of heavy metals that can easily be accumulated by the algae. Certain samples analyzed from such types of ponds have been observed to contain relatively higher levels of heavy metals (unpublished). Since certain plastic pond linings can also be sources of heavy metals, care should be taken during the selection of lining materials.

Table 8.5 shows the level of heavy metals in Earthrise Farms *Spirulina* (1984–1993) in relation to the very few guidelines available. The data show that the heavy metal concentrations are much lower than the guidelines for single-cell protein established by the United Nations Protein Advisory Group (UNPAG) (1974) and those established by Japan Health Food Association for *Chlorella* and *Spirulina*.

Pesticides

No pesticides are used during the cultivation of *Spirulina* at Earthrise Farms. However, since Earthrise Farms is located in an agricultural area there is a possibility

Table 8.5 Heavy-metal content of Earthrise Farms *Spirulina* in relation to international standards

	Lead	Mercury	Cadmium (ppm)	Arsenic	Total as lead
Aquatic animals (USFDA)		<1.0			
Single-cell protein (UNPAG)	<5.0	<0.1	<0.1	<2.0	
Chlorella (JHFA)				<2.0	<20
Spirulina (JHFA)				<2.0	<20
Spirulina (EF standard)	<1.0	<0.05	<0.05	<1.0	
Spirulina (EF 1988 & 1991)	<0.32	0.01	0.01	1.8	3.4

USFDA = United States Food and Drug Administration
UNPAG = United Nations Protein Advisory Group
JHFA = Japan Health Food Association
EF = Earthrise Farms

that the incoming water might contain pesticides. Even though the medium's high pH discourages the persistence of many pesticide compounds (Jassby, 1988a), it is imperative to monitor periodically for pesticides in the product. Periodic analysis of over thirty different pesticides in independent laboratories has failed to show any detectable levels of these compounds. Since routine analysis is expensive and periodic analysis alone cannot guarantee safety, Earthrise Farms employs a bioassay method that involves the stocking of fish in the reservoir that supplies the raw water. The condition of the fish in the reservoir is observed on a daily basis. *Spirulina* is known to be sensitive to many pesticides (Bednarz, 1981). As suggested by Jassby (1988a), an unusual productivity decline may be used as a warning signal for pesticide effects.

Cost of Production

Determinants of Cost of Production

The economic aspect of *Spirulina* production has been recently reviewed (Jassby, 1988b; Vonshak, 1992). The fixed costs of *Spirulina* production in terms of capital cost vary significantly with time. Jassby (1988b) has attempted to give estimates of fixed and variable costs based on the experiences of Earthrise Farms in pre-1988 conditions. In the latter review, Vonshak (1992) has compared the Earthrise data with those of Neotech in Bangkok. It is not intended to give an analysis of fixed costs here but merely to compare the current variable costs with those given by Jassby (1988b). It is interesting to note that there is no difference in the relative contribution of the various components of variable cost (Table 8.6) even though the pond area has been increased 1.5 times since his report. At Earthrise Farms, the major contributors to cost of production are labor, followed by operation, repair and maintenance costs.

As shown in Figure 8.10, the annual production per unit area in 1991 was about twice that of 1984. During the same period, the cost of production had gone down to less than 50 per cent of the 1984 value. Part of this reduction is attributable to economics of scale, since the total pond area was increased from 50 000 to 75 000 m². The main reasons for this significant reduction in cost of production since 1990 are (1) improved growth rates (Figure 8.11), (2) the ability to extend the growth season by one more month by a combination of strain selection and changes in culture conditions, (3) the ability to overwinter *Spirulina* in open ponds, thereby significantly reducing the time needed for culture expansion, (4) improved culture

Table 8.6 Relative cost of production

	1991	1992	1993	Average
Labor	38	36	38	37
Operations	29	29	30	29
Repairs & maintenance	5	4	5	5
Depreciation	15	15	15	15
Administrative Cost	12	16	13	14
Net Operating Cost	100	100	100	100

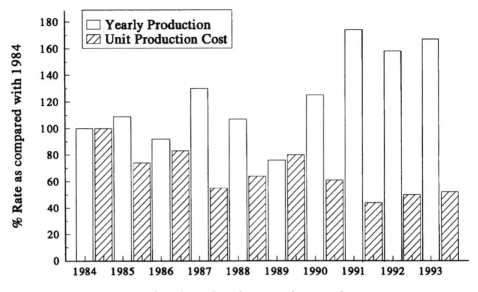

Figure 8.10 A comparison of yearly areal production and unit production cost (1984–1993).

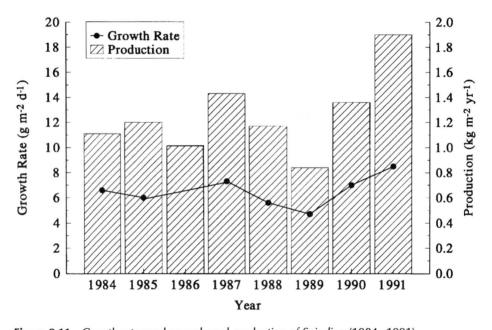

Figure 8.11 Growth rates and annual areal production of *Spirulina* (1984–1991).

and harvest conditions allowing higher harvest efficiencies and (5) increase in surface area for cultivation.

A significant reduction in cost of production can be achieved by selecting strains that are adapted to the colder temperatures of late winter and early spring, extending the duration of the production season by at least two months. Further optimization of culture conditions will no doubt result in still higher yields.

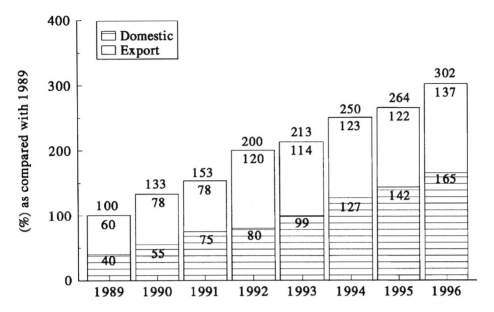

Figure 8.12   Past and future market demand for Earthrise Farms *Spirulina*.

Market Demand

The market demand for Earthrise Farms *Spirulina* is given in Figure 8.12. The values for 1989–1993 are actual demands while those for 1994–1996 are estimates. A steady increase is expected for domestic as well as export markets. The demand for Earthrise Farms *Spirulina* for both export and domestic consumption has doubled between 1989 and 1993. It is expected to rise by about 50 per cent between 1993 and 1996.

Conclusions

While there is a dearth of information on outdoor mass culture of *Spirulina* from small-scale and short-term studies, the published information from large-scale production systems is scanty and information from large-scale commercial production facilities is virtually absent. While some of the published information on outdoor mass culture of *Spirulina* derived from short-term and/or small-scale production systems is useful in elucidating some basic phenomena, it is not always applicable to the large-scale commercial production of *Spirulina*. There are problems encountered in large-scale systems which are rarely encountered in small-scale systems. Problems that are experienced in extended periods of operation are different, at least in magnitude, to problems met in short-term studies. Data from such short-term and/or small-scale system studies also do not take into consideration the economics of production, which are of high relevance to commercial production. Moreover, it is rarely that such small-scale studies, be they experimental or pilot plants, report the stability of the biochemical and microbiological characteristics of the final product.

155

The major commercial producers of *Spirulina* have, no doubt, a lot of accumulated knowledge on the art. However, because of business risk, not all the information that they have gathered through their long period of research and development is available for the general public. These companies have reached a stage of development in production know-how that new producers cannot attain without devoting some time and money. This realization may eventually result in more confidence by these successful producers to disclose more and more information on outdoor mass cultivation. Over the last few years, Earthrise Farms has been actively participating in international conferences in algal biotechnology. Through these participations it has been possible to exchange ideas in outdoor mass culture of *Spirulina*. We hope the other producers will eventually follow suit.

There are many researchers all over the world who are currently engaged in mass culture of *Spirulina*. Most of these researchers are affiliated with universities or research institutions where basic and applied research are done. They generate a lot of valuable information from laboratory and small-scale systems that can be tried out on large-scale facilities. Even within the constraints of free dissemination of information that prevail in commercial plants, a certain degree of research cooperation between industry and university or research organizations is possible. This will certainly benefit both parties. Researchers at Earthrise Farms have started cooperative research programs with other researchers in the United States and Japan in some areas of common interest. Each production facility may have its own specific problems that it may have to address in its own way. Some information may also be considered a trade secret for one reason or other. While some problems are unique to the specific production environment, there are other problems that are common to all production facilities. Cooperation among these facilities to solve common problems can significantly reduce the cost of research.

Acknowledgements

I am indebted to Mr Yoshimichi Ota, President of Earthrise Farms and Mr Hidenori Shimamatsu, Manager of Overseas Affairs, DIC, for their valuable contribution during the preparation of this manuscript.

References

ALDRICH, J. M. (1912) The biology of some western species of the dipterous genus *Ephydra*, *J. N. Y. Entomol. Soc.*, **20**, 77.

AOAC (1990a) *Official Methods of Analysis*, Fifteenth Edn, HELRICH, K., (Ed.) Association of Official Analytical Chemists, Inc., Arlington, 372.

AOAC (1990b) *Official Methods of Analysis*, Fifteenth Edn, First Supplement, HELRICH, K., (Ed.) Association of Official Analytical Chemists, Inc., Arlington, 17.

BAI, N. J. (1980) Competitive exclusion or morphological transformation? A case study with *Spirulina fusiformis*, Arch. Hydrobiol. Suppl. 71, *Algological Studies*, **38**, 191.

BAI, N. J. and SESHADRI, C. V. (1980) On coiling and uncoiling of trichomes in the genus *Spirulina*, Arch. Hydrobiol. Suppl. 60, *Algological Studies*, **26**, 32.

BECKER, E. W. (1994) *Microalgae: Biotechnology and Microbiology*, Cambridge Studies in Biotechnology 10, Cambridge: Cambridge University Press.

BEDNARZ, T. (1981) The effect of pesticides on the growth of green and blue-green algal cultures, *Acta Hydrobiol.*, **23**, 155.

BELAY, A. (1981) An experimental investigation of inhibition of phytoplankton photosynthesis at lake surfaces, *New Phytol.*, **89**, 67.

BELAY, A. (1993) Harvest-induced mechanical and physiological damage and recovery in *Spirulina*, in MASOJÍDEK, J. and SETLIK, I. (Eds) *Book of Abstracts of the 6th International Conference on Applied Algology*, Czech Republic, p. 48.

BELAY, A. and FOGG, G. E. (1978) Photoinhibition of photosynthesis in *Asterionella formosa* (Bacillariophyceae), *J. Phycol.*, **14**, 341.

BELAY, A. and OTA, Y. (1994) Temporal variations in the quality of *Spirulina* products from Earthrise Farms: a ten year study. Paper presented at *Second Asia-Pacific Conference on Algal Biotechnology*, Singapore, 25–27 April.

BELAY, A., OTA, Y., MIYAKAWA, K. and SHIMAMATSU, H. (1993) Current knowledge on potential health benefits of *Spirulina*, *J. Appl. Phycol.*, **5**, 235.

BONNIN, G. (1992) *B.E.C.C.M.A.'S* Spirulina *Production Engineering Handbook*, B.E.C.C.M.A., Nantes, 56.

CIFFERI, O. (1983) *Spirulina*, the edible microorganism, *Microbiol. Rev.*, **47**, 551.

DANGEARD, P. (1940) Sur une algue bleue alimentaire pour l'homme: *Arthrospira platensis* (Nordst.) *Gomont. Actes Soc. Linn. Boreaux Extr. Procés-verbaux*, **91**, 39.

EISENBERG, W. V. (1985) Sources of food contaminants, in GORHAM, J. R. (Ed.) *FDA Technical Bulletin*, Assoc. of Official Analytical Chemists in Cooperation with the FDA, Arlington, 15.

GOLDMAN, J. C. (1978) Outdoor algal mass cultures – II: Photosynthetic yield limitations. *Water Research*, **13**, 119.

JACQUET, J. (1976) Microflora of *Spirulina* preparations, *Ann. Nutr. Aliment.*, **29**, 589.

JASSBY, A. (1988a) Public health aspects of microalgal products. In LEMBI, C. A. and WAALAND, J. R. (Eds) *Algae and Human Affairs*, p. 182, Cambridge: Cambridge University Press.

JASSBY, A. (1988b) *Spirulina*: A model for microalgae as human food. In LEMBI, C. A. and WAALAND, J. R. (Eds) *Algae and Human Affairs*, p. 149, Cambridge: Cambridge University Press.

JAY, J. M. (1992) *Modern Food Microbiology*, p. 413, New York: Van Nostrand Reinhold.

LACQUERBE, B., BUSSON, F. and MAIGROT, M. (1970) On the mineral composition of two cyanophytes, *Spirulina platensis* (Gom) Geitler and *S. geitleri* J. de Toni, *C.R. Acad. Sci. Paris*. ser. D, **270**, 2130.

LEWIN, R. A. (1980) Uncoiled variants of *Spirulina platensis* (Cyanophyceae: Oscillatoriacea), *Arch. Hydrobiol. Suppl.* 60, *Algological Studies*, **26**, 48.

NAKASHIMA, M. J. (1989) Extraction of light filth from *Spirulina* powders and tablets: Collaborative study, *J. Assoc. Off. Anal. Chem.*, **72**, 451.

NEALE, P. J. (1987) Algal photoinhibition and photosynthesis in the aquatic environment. In KYLE, D. J., OSMOND, C. B. and ARNTZEN, C. J. (Eds) *Photoinhibition*, Amsterdam: Elsevier Science.

OSWALD, W. J. (1988) Large-scale algal culture systems (engineering aspects). In BOROWITZKA, M. A. and BOROWITZKA, L. J. (Eds) *Microalgal Biotechnology*, p. 357, Cambridge: Cambridge University Press.

RICHMOND, A. (1990) Large scale microalgal culture and applications. In ROUND, F. E. and CHAPMAN, D. J. (Eds) *Progress in Phycological Research*, Vol. 7, London: Biopress.

RICHMOND, A. and GROBBELAAR, J. U. (1986) Factors affecting the output rate of *Spirulina platensis* with reference to mass cultivation, *Biomass*, **10**, 253.

RICHMOND, A. and VONSHAK, A. (1978) *Spirulina* culture in Israel, *Arch. für Hydrobiol.*, **11**, 274.

RICHMOND, A., VONSHAK, A. and ARAD, S. M. (1980) Environmental limitations in outdoor production of algal biomass. In SHELEF, G. and SOEDER, C. J. (Eds) *Algal Biomass*, p. 65, Amsterdam: Elsevier/North-Holland Biomedical Press.

RICHMOND, A., LICHTENBERG, E., STAHL, B. and VONSHAK, A. (1990) Quantitative assessment of the major limitations on productivity of *Spirulina platensis* in open raceways, *J. Appl. Phycol.*, **2**, 195.

SHIMAMATSU, H. (1987) A pond for edible *Spirulina* production and its hydraulic studies, *Hydrobiologia*, **83**.

THEIN, M. (1993) Production of *Spirulina* in Myanmar (Burma). In DOUMENGE, F., DURAND-CHASTEL, H. and TOULEMONT, A. (Eds) *Bulletin de l'Institut Océanographique*, p. 175, Monaco: Musée Océanographique.

TORZILLO, G., SACCHI, A., MATERASSI, R. and RICHMOND, A. (1991) Effect of temperature on yield and night biomass loss in *Spirulina platensis* grown outdoors in tubular photobioreactors, *J. Appl. Phycol.*, **3**, 103.

UNPAG (1974) PAG guideline on nutritional safety aspects of novel protein sources for animal feeding, *PAG Bull.*, **15**, 11.

VENKATARAMAN, L. V. and KANYA, T. C. S. (1981) Insect contamination (*Ephydra californica*) in the mass outdoor cultures of blue green, *Spirulina platensis, Proc. Indian Acad. Sci. Sect.*, B, **90**, 665.

VONSHAK, A. (1987a) Biological limitations in developing the biotechnology for algal mass cultivation, *Science de L'Eau.*, **6**, 99.

VONSHAK, A. (1987b) Strain selection of *Spirulina* suitable for mass production, *Hydrobiologia*, **151**, 75.

VONSHAK, A. (1992) Microalgal biotechnology: is it an economical success? In DASILVA, E. J., RATLEDGE, C. and SASSON, A. (Eds) *Biotechnology: Economic and Social Aspects*, p. 70, Cambridge: Cambridge University Press.

VONSHAK, A. and GUY, R. (1988) Photoinhibition as a limiting factor in outdoor cultivation of *Spirulina platensis*. In STADLER, T., MOLLION, J., VERDUS, M. C., KARAMANOS, Y., MORVA, H. and CHRISTIAEN, D. (Eds) *Algal Biotechnology*, p. 365, Amsterdam: Elsevier Applied Science.

VONSHAK, A. and RICHMOND, A. (1988) Mass production of the blue-green alga *Spirulina*: an overview, *Biomass*, **15**, 233.

VONSHAK, A., ABELIOVICH, A., BOUSSIBA, S., ARAD, S. and RICHMOND, A. (1982) Production of *Spirulina* biomass: Effects of environmental factors and population density, *Biomass*, **2**, 175.

VONSHAK, A., BOUSSIBA, S., ABELIOVICH, A. and RICHMOND, A. (1983) Production of *Spirulina* biomass: Maintenance of pure culture outdoors, *Biotechnol. Bioeng.*, **25**, 341.

WIRTH, W. W. (1971) The brine flies of the genus *Ephydra* in North America (Diptera: Ephydridae), *Ann. Ent. Soc. Amer.*, **64**, 357.

ZAROUK, C. (1966) Contribution a l'étude d'une cyanophycée. Influence de divers facteurs physiques et chimiques sur la croissance et la photosynthese de *Spirulina maxima* (Setch. et Gardner) Geitler, Ph.D. thesis, University of Paris, France.

9

Mass Cultivation and Wastewater Treatment Using *Spirulina*

GILLES LALIBERTÉ, EUGENIA J. OLGUIN AND JOËL DE LA NOÜE

Introduction

In the past fifteen years, the use of microalgae and cyanobacteria for wastewater treatment has been reviewed by many authors (Laliberté et al., 1994; de la Noüe et al., 1992; Lincoln and Earle, 1990; Oswald, 1988a, 1988b; 1991). These reviews demonstrate that algal cultures in high-rate oxidation ponds (HROP) for the removal of nitrogen and phosphorus in parallel with the production of a useful algal biomass can be an interesting alternative to conventional tertiary wastewater treatment. The most abundant algae found in the water of naturally managed HROPs are generally of the genera *Chlorella*, *Ankistrodesmus* and *Scenedesmus*. However, the removal of these small algae from the effluent can represent a major cost associated with algal cultivation on wastewater (Mohn, 1988). Furthermore, because well designed HROP can produce more than 200 kg dry mass of algae per hectare per day, disposal of the concentrated algal biomass requires convenient solutions.

Within that context, wastewater treatment with *Spirulina* offers many advantages: (1) its filamentous form and its capacity to bioflocculate makes its harvesting easier and less costly than for other microalgae (Mohn, 1988); (2) under appropriate conditions, its protein content can attain 60 to 70 per cent of its dry mass (Ciferri, 1983); (3) its nutritional value as animal feed is recognized (Jassby, 1988; Kay, 1991; Becker, 1994; Shubert, 1988); (4) it contains many biochemical compounds which can be extracted for potential enhancement of the biomass value (Belay et al., 1993); (5) it grows best in an alkaline medium; and (6) some species are resistant to ammonia toxicity found at high pH values (Belkin and Boussiba, 1991).

Treatment and Recycling of Animal Wastes

In order to use *Spirulina* grown on wastewater as animal feed, it is necessary to minimize contamination of the biomass. This can be achieved by choosing the right type of effluent to be treated. For example, it is obvious that wastewaters from industrial processes, containing high amounts of heavy metals, are unsuitable for the cultivation of algae as a source of animal feed. In that respect, animal wastes represent a good source of substrates for the culture of *Spirulina*. By employing an integrated approach one can easily conceive the treatment of wastes with concomitant production of algal biomass for animal feeds. This is probably why most of the research reported on the treatment of wastewater with *Spirulina* deals with the recycling of animal wastes and most often with pig wastes, because intensive pig production is causing very serious problems of water pollution worldwide.

Animal Wastes

While municipal and industrial wastewaters are liquid wastes containing some solids, animal wastes in general can be considered as solid matters containing some water (Andreadakis, 1992). From the typical composition of various animal wastes (Table 9.1), one can see that poultry, sheep and pig wastes contain the highest load of nitrogen and phosphorus (Newell, 1980; Sweeten, 1991), two nutrients required for the growth of microalgae. Indeed, in the 1970s, the production of microalgae (*Chlorella, Scenedesmus*) on fresh manure was attempted (Boersma et al., 1975). However, because at least a 50-fold dilution of the manure was necessary for the establishment of an algal population, and a high rate of ammonia volatilization was observed, the impracticality of such a system was concluded (Boersma et al., 1975; Shuler, 1980).

A typical medium suitable for the growth of *Spirulina* contains higher amounts of bicarbonate ($13.6 \text{ g} \text{l}^{-1}$), nitrate ($2.5 \text{ g} \text{l}^{-1}$ vs 0.005) (de la Noüe and Bassères, 1989), potassium ($1.0 \text{ g} \text{l}^{-1}$ vs 0.4), and sodium chloride ($1.0 \text{ g} \text{l}^{-1}$ vs 0.1) (Wu and Pond, 1981) than those typically found in pig wastes. Although it is possible to grow *Spirulina* in 35-fold diluted raw pig waste by adding bicarbonate ($4 \text{ g} \text{l}^{-1}$) and sodium chloride ($4 \text{ g} \text{l}^{-1}$) (Pouliot et al., 1986), it is not advantageous to do so. First of all, as stated before, such a large dilution makes this method economically impractical. Secondly, the extreme variability in the composition of pig waste from one batch to another leads to variable growth rates of the alga and makes it difficult to optimize the system. Thirdly, aerobic or anaerobic fermentation of the wastes is necessary in order to lower their biological oxygen demand. Fourthly, anaerobic digestion of the wastes becomes an economically attractive method if methane and carbon dioxide productions are maximized. In fact, by comparing the main

Table 9.1 Fresh manure production and characteristics (Newell, 1980; Sweeten, 1991)

Constituent	Cow	Beef	Swine	Sheep	Poultry
Manure (kg per day per 454 kg of live weight)	39	26	38	18	29
Total solids (%)	14	15	13	28	25
Volatile solids (%)	12	12	10	23	18
N (kg per tonne of waste)	3.8–5.2	5.0–5.9	5.4–6.3	11.3	8.8–13.1
P (kg per tonne)	1.1–1.8	1.6–2.5	2.2–3.1	2.1	4.6–8.2
K (kg per tonne)	3.4–4.8	3.6–4.5	3.5–4.1	8.0	3.2–4.8
BOD range ($g \text{ l}^{-1}$)	6–50	6.7–50	12–60	—	10–800

advantages and disadvantages of anaerobic digestion with those of aerobic digestion (Table 9.2), it is quite obvious that in countries limited in energy or where lower operation costs are required, anaerobic digestion is the procedure of choice for the treatment of animal wastes.

Cultivation of Spirulina *on animal wastes*

Representative data published on the cultivation of *Spirulina* on animal wastes are summarized in Table 9.3. A close inspection of these data leads to the following conclusions: (1) anaerobic digestion of the waste is the usual practice; (2) a high dilution (in the range 90 to 98 per cent) of the digested waste is necessary; (3) the temperature used is about 30 °C; (4) for optimum growth, the addition of sodium bicarbonate is required. Furthermore, it should be noted that many investigations have been performed at laboratory scale in very small volumes. Because extrapolation from small cultures is often very unreliable (Borowitzka, 1992), it is difficult to predict the behavior of such systems in large open ponds.

Anaerobic or aerobic digestion of the wastes

Although most research groups have recognized the importance of integrating primary and secondary treatments of animal wastes with the cultivation of *Spirulina*, little attention has been paid to the optimization of these treatments in order to provide effluents with optimum composition for *Spirulina* cultivation. In general, anaerobic treatment increases the pH of the influent without changing its ammonium concentration, resulting in a medium better suited to the growth of cyanobacteria than that from an aerobic treatment (Table 9.4). In fact, renewed interest in anaerobic digestion has given rise to reactors easier to control under farm conditions and with shorter retention times. It will be of great interest to adapt these reactors to furnish an effluent suitable for the growth of cyanobacteria. For example, Olguin et al. (1994b) reported a solid loading rate of 3 kg volatile solid $m^{-3} d^{-1}$ for a mesophilic plug-flow reactor treating pig waste and giving an effluent suitable for the growth of *S. maxima*. In addition, under anaerobic digestion, animal wastes have the potential to produce about 0.3 m^3 of biogas per kg of total solids, containing about 70 per cent methane (Dodson and Newman, 1980).

Table 9.2 Advantages and disadvantages of anaerobic and aerobic digestion

Digestion	Advantages	Disadvantages
Anaerobic	Production of methane; Lower amount of activated sludge produced; Low capital investment with second generation reactors	Sensitive to operating conditions (flux, BOD, pH, T); More or less stable; Less suitable for complex effluents
Aerobic	Easier to operate; More stable; Low capital investment; Easier to recycle the sludge	Energy consuming; No production of methane; Production of sludge in high quantity

Table 9.3 Examples of environmental conditions used for the cultivation of *Spirulina* on animal wastes

Species	Type of waste	Volume (l)	Temp. (°C)	Light	HCO_3^- (g l⁻¹)	NaCl (g l⁻¹)	$N-NH_3$ (mg l⁻¹)	PO_4^{-3} (mg l⁻¹)	Reference
platensis	pig anaerobic 10% v/v	174	33	500 fc	1.7–16.8	0.5	185	425	Chung et al., 1978
maxima	cow 1% v/v	200	30	outdoors	16	—	—	127	Oron et al., 1979
platensis	pig anaerobic	10	30	outdoors	1–3	3–7	100	—	Chiu et al., 1980
maxima	pig anerobic 2% v/v	1	30	80 $\mu E\ m^{-2}\ s^{-1}$	2	saline water	—	—	Olguin and Vigueras, 1981
platensis	poultry anaerobic 2% v/v	2000	30	outdoors	4	0.5	~30	128	Venkataraman et al. 1982; Mahadevaswamy and Venkataraman, 1986
fusiformis	cattle	—	30	outdoors	8	1	—	250	Thomas, 1982
maxima	pig–cattle anaerobic 3–4.5% v/v	300	22	3–30 lux	1–2	5–7 sea salt	100–1200	100–250	Ayala and Bravo, 1984; Ayala and Vargas, 1987
sp.	pig anaerobic 0.8% v/v	12500	—	outdoors	8.4	0.5	—	250	Yang and Duerr, 1987
platensis	pig anaerobic 2% v/v	1	25	—	10	—	465	1.7	Gantar et al., 1991
maxima	pig aerobic 2% v/v	2	room temp.	1500 lux	8.40	.5	83	4	Canizares and Dominguez, 1993
maxima	pig anaerobic 2% v/v	20	30	70 $\mu E\ m^{-2}\ s^{-1}$	4.0	sea water	25.2	13.6	Olguin et al., 1994a
platensis	cattle anaerobic	0.5	36	34 lux	2–4	0.3	17	49	Fedler et al., 1993

Table 9.4 Evolution of some parameters following anaerobic or aerobic digestion of pig waste (Olguin et al., 1994a; Martinez and Burton, 1994)

Parameter	Anaerobic	Aerobic
	$\left(\dfrac{\text{value in the effluent}}{\substack{\text{value in the influent} \\ \text{(except for pH)}}} \times 100\right)$	
Total solids	14	85
Volatile solids	12	0.5
Total N	73	66
N–NH$_3$	106	22
P–PO$_4^{3-}$	35	—
pH	5.1 →9.3	7.2→7.1

Constraints

The cultivation of *Spirulina* on digested waste encounters the same constraints as those of commercial algal production, as for example the availability of carbon, nitrogen and phosporus, optimum temperature and light conditions, as well as photoinhibition (Richmond, 1990). While the amount of nitrogen and phosphorus might be sufficient in these media for the growth of *Spirulina*, bicarbonate and sodium chloride still need to be added. When available, commercial sodium chloride can be successfully replaced by sea water for the growth of this cyanobacterium (Olguin et al., 1994b). Interestingly, while natural sources of salt water are a major problem for the development of agriculture in many southwestern areas of the United States, Parker et al. (1992) advocate the use of these brine springs for the growth of *Spirulina* on anaerobically digested cattle wastes. They also point out that *Spirulina* can grow in diluted oilfield brines following bacterial degradation of their residual hydrocarbon.

Bicarbonate Needs

Among all the costs associated with the culture of *Spirulina* (capital, labor, water, NaCl, nutrients and power) it has been calculated that carbon represents one of the major components in the operating cost (Borowitzka, 1992). While early investigators added sodium bicarbonate at concentrations of 9 to 17 g l^{-1} for the growth of *Spirulina* on digested wastes, some works show that bicarbonate addition in the range of 3 to 4 g l^{-1} is sufficient (Fedler et al., 1993; Becker and Venkataraman, 1982). The use of the CO$_2$ produced from anaerobic digesters has also been proposed as a means to lower the cost associated with the supply of inorganic carbon (Venkataraman et al., 1982), but this can favor contamination with *Chlorella* species (Ip et al., 1982; Richmond et al., 1982). In fact, the high bicarbonate and pH conditions needed for the growth of *Spirulina* form a good barrier against contamination by other microalgae. It seems that with their high affinity for CO$_2$ and their capacity to utilize bicarbonate, cyanobacteria outcompete eukaryotic algae in conditions of low carbon dioxide concentration and high pH value (Shapiro, 1990). In addition, while *S. platensis* can grow in heterotrophy and mixotrophy on 2 gl^{-1} of glucose (Marquez et al., 1993), this phenomenon is likely to be of little importance for the cultivation of *Spirulina* on wastes. On the contrary, it has been demonstrated that excessive accumulation of organic matter in outdoor cultures of *Spirulina* could inhibit its growth (Belay et al., 1994) and lead to contamination by other algae. Similarly, Gantar et al. (1991) demonstrated that *S. platensis* was displaced by autochthonous microalgae when the concentration of swine waste was increased in the culture medium. However, because of its potential implication, the role of heterotrophy and mixotrophy during *Spirulina* growth should be further investigated.

Depollution or Biomass Production

Ideally, biological treatment of wastes with cyanobacterial cultures should achieve maximal removal rates of nutrients together with a high biomass production. However, in reality, one has to emphasize either depollution or biomass production because there is no obligatory correlation between these two parameters (de la Noüe and Bassères, 1989). In fact, the ideal operational conditions necessary to obtain a high rate of depollution often lead to low biomass output rate, and vice versa.

Removal of Inorganic Nutrients

Surprisingly, while many investigators have emphasized the cultivation of *Spirulina* on digested wastes to remove inorganic nutrients, very little reliable data exist on nutrient removal efficiency by this cyanobacterium. Rates of nitrogen (10–18 mg$l^{-1}d^{-1}$) and phosphorus (1.0–1.5 mg$l^{-1}d^{-1}$) removal reported in the literature (Pouliot et al., 1986), during the growth of *Spirulina* on animal wastes agree well with those reported with other cyanobacteria such as *Phormidium bohneri* grown on anaerobically digested swine manure (de la Noüe and Bassères, 1989). While there is no doubt of the efficiency of such systems, one has to keep in mind that much of it can be due to abiotic, rather than biotic factors. With this type of process, nitrogen is mainly removed by active uptake during cyanobacterial growth or by ammonia-stripping into the atmosphere, since the main form of nitrogen found in animal wastes is the ammonium ion. The optimal conditions needed for the growth of *Spirulina*, that is high temperature and pH values, and a good aeration rate, also greatly favor the stripping of ammonia. Unfortunately, the loss of gaseous NH_3 during the cultivation of *Spirulina* has rarely been measured in an outdoor system. However, knowing that ammonia stripping accounted for at least 62 per cent of nitrogen removal in a 75 l outdoor culture of the cyanobacterium *Phormidium bohneri* (Proulx et al., 1994) and that a good correlation has been found between *in situ* rates of ammonia removal in waste stabilization ponds and those predicted by an ammonia-stripping model (Pano and Middlebrooks, 1982), one can conclude that this phenomenon is a major factor for the removal of ammonia in *Spirulina* culture. Because gaseous ammonia is also a pollutant, there is an urgent need to quantify rates of ammonia volatilization during cyanobacterial growth to see if, at least in terms of nitrogen, one does not simply transfer the pollution problem from water to air. On the other hand, in the presence of sodium chloride, ammonia volatilization is diminished and more nitrogen remains available for algal assimilation (Chiu et al., 1980). It would be of interest to investigate the relationship between the amount of sodium chloride added, the growth rate of *Spirulina* and the degree of ammonia stripping.

The removal of inorganic phosphorus in algal culture results from two phenomena: biological assimilation and chemical precipitation as insoluble phosphate. As with ammonia nitrogen, the environmental conditions that prevail during *Spirulina* cultivation also favor the abiotic process of phosphate removal. Again, this phenomenon has never been thoroughly investigated and needs to be seriously studied in order to establish a typical phosphorus balance in cultures of cyanobacteria.

Ammonia Toxicity

One drawback of the cultivation of *Spirulina* on wastes, especially if the emphasis is put on the efficiency of inorganic nutrient removal, is the high dilution of the anaerobically digested waste needed for the growth of *Spirulina*. While this problem has never been thoroughly investigated, the toxicity of ammonia (NH_3) is certainly a major factor explaining the need for considerable dilution of the waste (Chiu et al., 1980). The predominance of NH_3 over NH_4^+ at pH values greater than 9.25 (the pK_a of the ammonia system at 25 °C), combined with the high permeability of the uncharged form of the molecule, causes uncoupling of photosynthesis in algal cells (Boussiba and Gibson, 1991). At pH values of 9.0 to 9.5, which are considered optimal for the growth of *Spirulina*, concentration of free ammonia above 2.5 mM is reported to be toxic to many algae (Abeliovich and Azov, 1976). On the other hand, *S. platensis*, by maintaining higher internal pH values than those of other cyanobacteria, was shown to be quite resistant to ammonia-uncoupling of photosynthesis (Belkin and Boussiba, 1991). In that context, it would certainly be advantageous to work with *Spirulina* strains resistant to ammonia toxicity in order to treat less diluted wastes. Furthermore, because ammonia concentration is usually lower in animal wastes following aerobic, as compared with anaerobic, digestion (Table 9.4), the volume of aerobically digested waste that can be treated by *Spirulina* cultivation is potentially much higher for a given surface area. For example, at laboratory scale, a maximal removal of nitrogen and phosphorus by *S. maxima* occurred in 50 per cent diluted swine waste stabilized by aeration, corresponding to about 42 mg nitrogen l^{-1} of nitrogen (Canizares and Dominguez, 1993).

Biomass Production

When cultivating *Spirulina* on animal wastes, the emphasis can also be put on biomass production rather than on epuration. In most systems investigated to date, a rate of algal production of about 10 g dry mass $m^{-2} d^{-1}$ has been reported (Chiu et al., 1980; Venkataraman et al., 1982; Yang and Duerr, 1987). However, much research needs to be done to optimize yield and quality of the biomass as a function of environmental factors such as light intensity and temperature. For example, Olguin et al. (1994a) reported that at low light intensity, a flow rate of 14.5 cm s^{-1} was optimum for accumulation of proteins in *S. maxima* cultivated in raceways containing anaerobic effluents from pig waste enriched with sea water. Higher flow rates at the same light intensity resulted in a higher biomass production but with a lower protein content. In outdoor cultures, photoinhibition often results in decreased productivity of *Spirulina* (Vonshak et al., 1988). This problem might be alleviated in animal wastes because they are often dark in color.

Innocuity of the Algal Biomass

While the safety of commercially produced *Spirulina* biomass as a source of protein for animals has been demonstrated in numerous studies (Kay, 1991; Shubert, 1988), the same thing cannot be said of *Spirulina* produced on wastewater. However, a few studies did show that the biomass of *Spirulina* produced on digested animal wastes or on domestic raw sewage, and sun-dried, has the potential to be used as fish or poultry feeds (Thomas, 1982; Saxena et al., 1982, 1983; Fox, 1988). More long-term studies are required in order to make sure that an algal biomass produced on wastes is indeed safe for animal consumption. *Spirulina* biomass as a source of protein for swine production is advantageous because a wet feed is possible and the costs of drying the biomass are thus much reduced (Benemann, 1989).

Treatment of Other Types of Wastes

Domestic or Industrial Wastewaters

Domestic wastewater

While studies exist about the use of microalgae such as *Chlorella* (Tam and Wong, 1990), *Scenedesmus* (Becker, 1994; Becker and Venkataraman, 1982) and of cyanobacteria such as *Phormidium* (Proulx et al., 1994) and *Oscillatoria* (Hashimoto and Furukawa, 1989) for the tertiary treatment of domestic wastewater, very few studies have been published on this subject with *Spirulina*. The first study showing the potential of growth of *Spirulina* on domestic wastewaters was published in 1974 by Kosaric et al. (1974). In that study, *S. maxima* was grown in 27 l containers on secondary effluents from domestic wastewater. The addition of 1 gl^{-1} of sodium bicarbonate was necessary in order to adjust the pH of the effluent, originally near neutrality, to 9.5. Beginning with concentrations of 52 ppm of nitrogen and 4.2 ppm of phosphorus, 87 per cent of the nitrogen and 60 per cent of the phosphorus were removed in 4 days. In 1983, Saxena et al. published a study of an integrated system for the growth of *Spirulina* in domestic raw sewage for poultry feed in India. They showed that *Spirulina* could adequately grow in raw sewage only if supplemented with 10 gl^{-1} of sodium bicarbonate and 1 gl^{-1} of sodium nitrate. Under these conditions, they obtained a yield of about 9 $g\,m^{-2}\,d^{-1}$ during a 12-day period. Interestingly, the biomass thus produced was not pathogenic to chicken when used as poultry feed and was a good yolk pigmenter (Saxena et al., 1982, 1983).

Palm oil mill effluent

More recently, Siew-Moi (1987) reported that *S. platensis* can grow in anaerobically treated palm oil mill effluent. With a detention time of 5 days, the maximum algal productivity was 33.8 g d.w. $m^{-2}\,d^{-1}$ and the resulting treated water was good enough to be recycled for factory use.

Starch wastewater from tapioca factory

The occurrence of *Spirulina* in stabilization ponds of secondary treated starch wastewater from tapioca factories in Thailand has also been reported (Tanticharoen et al., 1993). The authors isolated a *Spirulina* strain able to grow in the water from these stabilization ponds with the addition of bicarbonate and fertilizers. After sun-drying, the algal biomass, containing 55 per cent protein and 7 per cent moisture, was sold and used in the formulation of prawn and ornamental fish feed. In addition, Tanticharoen et al. (1993) suggested the integration of anaerobic digestion of wastewater from starch factories in Thailand with the cultivation of *Spirulina* in order to reduce the cost of production of this cyanobacterium.

It is worth noting that, based on this study, a commercial set-up is now under operation by a company called 'Neotech Food Co. Ltd' which is using the tapioca wastewater and has the capacity to produce up to 30 tonnes of biomass per year.

Human Wastes

One very important area of research related to the growth of *Spirulina* on wastewater is the development of uncomplicated and affordable technology designed to be used at the village scale in developing countries, in order to produce *Spirulina* for feeding animals or humans. In that respect, the pioneer works of L. V. Venkataraman and R. D. Fox have to be underlined. While only a quick overview can be given here, the extensive works of both authors have been summarized in three interesting books (Becker, 1994; Becker and Venkataraman, 1982; Fox, 1986). Venkataraman and collaborators have investigated various aspects of the production of *Scenedesmus* and *Spirulina* in the climatic and economic context of India. They showed, for example, that bone meal could be substituted for calcium and phosphate salts in *Spirulina* cultures and that human or cow urine at 1 per cent level could successfully replace sodium nitrate as a nitrogen source (Becker and Venkataraman, 1982). The biomass produced, with yields in the range of $8-12$ g dry mass $m^{-2} d^{-1}$, was collected by gravity and sun-dried. Results showed that such biomass grown on wastes can safely be incorporated at a 5 per cent level in chicken rations. On the other hand, in the last thirty years, Fox (1986) has spent a lot of time and energy on the development of a system integrating sanitation, biogas production, *Spirulina* cultivation, composting and fish culture in developing country villages. By using, as far as possible, local resources, Fox and his collaborators constructed three projects in India, Senegal and Togo (Fox, 1987, 1988). In all cases, liquid effluents from anaerobically digested human wastes are passed through a sand bed filter and heated in a solar sterilizer. These operations destroy almost 90 per cent of the pathogens present in the wastes. With the addition of 5 $g l^{-1}$ of ocean salt and 4 $g l^{-1}$ sodium bicarbonate, the sterilized effluent produced is used for the growth of *Spirulina* with a yield of 15 g dry mass $m^{-2} d^{-1}$. After solar drying and sterilization, this biomass has been used, with very encouraging results, as a source of additional protein for babies suffering severe malnutrition (Fox, 1987). In addition, the methane gas produced during anaerobic digestion is used as cooking fuel, some algal biomass is sold for fish culture and the digester sludge is converted into compost. There is no doubt that *Spirulina* farming projects should be encouraged and, as stated by Fox (1988), the elimination or reduction of malnutrition in developing country villages far outweighs the cost asssociated with this technology.

Effluents Containing Heavy Metals

The removal of heavy metals from polluted effluents is certainly a promising field for the use of microalgal and cyanobacterial biomass produced on wastewater. Recently, Wilde and Benemann (1993) concluded that biosorption of heavy metals has the potential to be more efficient and cheaper than that of quaternary wastewater treatment processes. *Spirulina*, among other algae such as *Chlorella, Scenedesmus, Chlamydomonas* and *Oscillatoria*, has been used for the removal of heavy metals (Bedell and Darnall, 1990). For example, alginate and polyacrylamide-immobilized *Chlorella vulgaris* and *Spirulina platensis* have been used to remove many metals, including Cu, Pb, Zn and Au. A commercially available product to remove heavy metals, in batch or in a column, called 'AlgaSORB™' from Bio-Recovery Systems, Inc., Las Cruces, New Mexico, USA, contains dead algal biomass, including *Spirulina* and *Chlorella*, immobilized in a silica matrix. With the appropriate treatment, it is possible to regenerate the biomass of AlgaSORB to about 90 per cent of the original metal-uptake efficiency, even after extended use (Gadd and White, 1993). Finally, a biosorbent using immobilized biomass in polysulfone, including *Spirulina*, yeast, microalgae and higher plants (*Lemna* sp., *Sphagnum* sp.) is available under the name of BIO-FIX. This material has decreasing affinity for $Al > Cd > Zn > Mn$, and can thus be used for selective removal of heavy metals. In fact, the capacity of algae to bioaccumulate heavy metals can sometimes be a problem for the cultivation of *Spirulina* in pig wastes. If the concentration of copper ion, originally added to the pigs' feed, in the anaerobic effluent of such wastes is too high, it quickly leads to cell mortality.

Conclusions

All of the above studies clearly show the potential of using *Spirulina* for wastewater treatment. However, an integrated system for biomass production and wastewater depollution as schematized in Figure 9.1, is indeed a very ambitious project. Much work needs to be done, both in the laboratory and in outdoor conditions, in order to reach such a goal. Among other things, there is a need to:

- adjust operating conditions of second-generation anaerobic digesters to the needs of *Spirulina* cultivation;
- isolate *Spirulina* strains resistant to ammonia toxicity;
- quantify the fate of nitrogen and phosphorus in *Spirulina* cultures;
- further characterize the safety of *Spirulina* biomass produced on wastes and used as animal or human feeds; although *Spirulina* itself is not harmful, there is always the danger of collecting toxic strains of cyanobacteria with that of *Spirulina* (Carmichael, 1994);
- investigate the capacity of *Spirulina* to grow on other types of effluents such as those demonstrated with other cyanobacteria, as for example in fish farm effluent (Dumas et al., 1994).

There is no doubt that with some projects presently going on, for example, in Mexico (Olguin et al., 1994a, 1994b), in the United States (Fedler et al., 1993) and in Thailand (Tanticharoen et al., 1993), the future development of wastewater

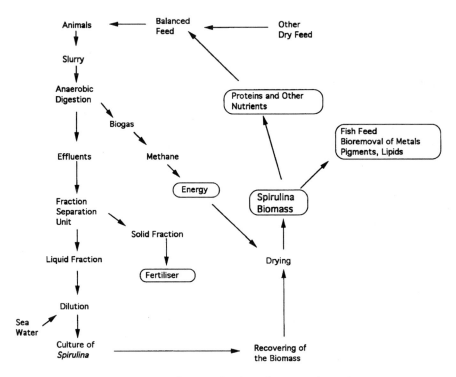

Figure 9.1 Example of an integrated system for the cultivation of *Spirulina*.

treatment with *Spirulina* looks very promising. However, such systems will have to form a unified whole, as shown in Figure 9.1, to be economically viable. Since raw materials for *Spirulina* production in integrated systems are often animal wastes and sea water, an algal biomass much cheaper than that produced on nitrate, NaCl and fresh water, is expected. If the biomass thus produced is then recycled via animal and fish feeds, used as an inert support for bioremoval of heavy metals, or used as a fuel, the whole system could indeed become viable.

References

ABELIOVICH, A. and AZOV, Y. (1976) Toxicity of ammonia to algae in sewage oxidation ponds, *Appl. Envir. Microbiol.*, **31**, 801.

ANDREADAKIS, A. D. (1992) Anaerobic digestion of piggery wastes, *Water. Sci. Technol.* **25**, 9.

AYALA, F. J. and BRAVO, R. B. (1984) Animal wastes media for *Spirulina* production. *Arch. Hydrobiol.*, **67**, 349.

AYALA, F. J. and VARGAS, T. (1987) Experiments on *Spirulina* culture on waste-effluent media and at the pilot plant. In RAGAN, M. A. and BIRD, C. J. (Eds) Twelfth-International Seaweed Symposium, Sao Paulo, Brazil, *Hydrobiologia*, **151–152**, 91.

BECKER, E. W. (1994) *Microalgae: Biotechnology and Microbiology*, Cambridge: Cambridge University Press.

BECKER, E. W. and VENKATARAMAN, L. V. (1982) *Biotechnology and Exploitation of Algae: The Indian Approach*, Deutsche Gesellschaft für Technische Zusammenarbeit (GTZ), Germany.

BEDELL, G. W. and DARNALL, D. W. (1990) Immobilization of non-viable, biosorbent, algal biomass for the recovery of metal ions. In VOLESKY, B. (Ed.) *Biosorbents and Biosorption Recovery of Heavy Metal Ions*, 313, Boca Raton, Fl.: CRC Press.

BELAY, A., OTA, Y., MIYAKAWA, K. and SHIMAMATSU, H. (1993) Current knowledge on potential health benefits of *Spirulina, J. Appl. Phycol.*, **5**, 235.

BELAY, A., OTA, Y., MIYAKAWA, K. and SHIMAMATSU, H. (1994) Production of high quality *Spirulina* at Earthrise farms. In SIEW MOI, P., YUAN KUN, L., BOROWITZKA, M. A. and WHITTON, B. A. (Eds) *Algal Biotechnology in the Asia-Pacific Region*, p. 92, Malaysia: Institute of Advanced Studies, University of Malaya.

BELKIN, S. and BOUSSIBA, S. (1991) Resistance of *Spirulina platensis* to ammonia at high pH values, *Plant Cell Physiol.*, **32**, 953.

BENEMANN, J. R. (1989) The future of microalgal biotechnology. In CRESSWELL, R. C., REES, T. A. V. and SHAH, N. (Eds) *Algal and Cyanobacterial Biotechnology*, 317, Longman Scientific Technical, UK.

BOERSMA, L., BARLOW, E. W. R., MINER, J. R., PHINNEY, H. K. and OLDFIED, J. E. (1975) Protein production rates by algae using swine manure as substrate. In JEWELL, W. J. (Ed.) *Energy, Agriculture and Waste Management*, p. 475, Ann Arbor, Mich: Ann Arbor Science.

BOROWITZKA, M. A. (1992) Algal biotechnology products and processes-matching science and economics, *J. Appl. Phycol.*, **267**.

BOUSSIBA, S. and GIBSON, J. (1991) Ammonia translocation in cyanobacteria, *FEMS Microbiol. Rev.*, **88**, 1.

CANIZARES, R. O. and DOMINGUEZ, A. R. (1993) Growth of *Spirulina maxima* on swine waste, *Bioresource Technol.*, **45**, 73.

CARMICHAEL, W. W. (1994) The toxins of cyanobacteria, *Scientific American*, January, 78.

CHIU, R. J., LIU, H. I., CHEN, C. C., CHI, Y. C., SHAO, H., SOONG, P. and HAO, P. L. C. (1980) The cultivation of *Spirulina platensis* on fermented swine manure. In CHANG PO (Ed.) *Animal Wastes Treatment and Utilization*, Proc. Int. Symp. on Biogas, Microalgae and Livestock, Taiwan, p. 435.

CHUNG, P., POND, W. G., KINGSBURY, J. M., WALKER, E. F. and KROOK, L. (1978) Production of nutritive value of *Arthrospira platensis*, a spiral blue-green alga grown on swine wastes, *J. Animal Sciences*, **47**, 319.

CIFERRI, O. (1983) *Spirulina*, the edible microorganism, *Microbiol. Rev.*, **47**, 551.

DE LA NOÜE, J. and BASSÈRES, A. (1989) Biotreatment of anaerobically digested swine manure with microalgae. *Biol. Wastes*, **29**, 17.

DE LA NOÜE, J., LALIBERTÉ, G. and PROULX, D. (1992) Algae and waste water, *J. Appl. Phycol.* **4**, 247.

DODSON, C. and NEWMAN, T. (1980) Energy from biomass conversion systems. In VOGT, F. (Ed.) *Energy Conservation and Use of Renewable Energies in the Bio-Industries*, p. 151, Oxford: Pergamon Press.

DUMAS, A., DE LA NOÜE, J. and LESSARD, P. (1994) Biotreatment of a fish farm effluent using a cyanobacterium, *Phormidium bohneri*. In LALIBERTÉ, G. and DE LA NOÜE, J. (Eds) *Microalgae: From the Laboratory to the Field*, p. 99, Ste-Foy, Canada: Department of Food Science and Technology, Université Laval.

FEDLER, C. B., PULLUOGLU, M. A. and PARKER, N. C. (1993) Integrating livestock waste recycling with production of microalgae. In *Techniques for Modern Aquaculture*, American Society of Agricultural Engineers, Publication 02-93.

FOX, R. D. (1986) *Algoculture: La Spirulina, un Espoir pour le Monde de la Faim*, Edisud, La Calade, 13090 Aix-en-Provence.

FOX, R. D. (1987) *Spirulina*, real aid to development, *Hydrobiologia*, **151/152**, 95.

FOX, R. D. (1988) Nutrient preparation and low cost basin construction for village production of *Spirulina*. In STADLER, T., MOLLION, J., VERDUS, M.-C., KARAMANOS, Y.,

MORVAN, H. and CHRISTIAEN, D. (Eds) *Algal Biotechnology*, p. 355, New York: Elsevier Applied Science.

GADD, G. M. and WHITE, C. (1993) Microbial treatment of metal pollution – a working biotechnology? *Tibtech* , **11**, 353.

GANTAR, M., OBREHT, Z. and DALMACIJA, B. (1991) Nutrient removal and algal succession during the growth of *Spirulina platensis* and *Scenedesmus quadricauda* on swine wastewater, *Bioresource Technol.*, **36**, 167.

HASHIMOTO, S. and FURUKAWA, K. (1989) Nutrient removal from secondary effluent by filamentous algae, *J. Ferment. Bioeng.*, **67**, 62.

IP, S. Y., BRIDGER, J. S., CHIN, C. T., MARTIN, W. R. B. and RAPER, W. G. C. (1982) Algal growth in primary settled sewage: The effects of five key variables, *Water Res.*, **16**, 621.

JASSBY, A. (1988) *Spirulina*: a model for microalgae as human food. In LEMBI, C. A. and WAALAND, J. R. (Eds) *Algae and Human Affairs*, p. 149, Cambridge: Cambridge University Press.

KAY, R. A. (1991) Microalgae as food and supplement, *Crit. Rev. Food Sci. Nutr.*, **30**, 555.

KOSARIC, N., NGUYEN, H. T. and BERGOUGNOU, M. A. (1974) Growth of *Spirulina maxima* algae in effluents from secondary waste-water treatment plants, *Biotechnol. Bioeng.*, **16**, 881.

LALIBERTÉ, G., PROULX, D., DE PAUW, N. and DE LA NOÜE, J. (1994) Algal technology in wastewater treatment. In RAI, L. C., GAUR, J. P. and SOEDER, C. J. (Eds) *Algae and Water Pollution*, Archiv. for Hydrobiol., Advances in Limnol., **42**, 283.

LINCOLN, E. P. and EARLE, J. F. K. (1990) Wastewater treatment with microalgae. In AKATSUKA (Ed.) *Introduction to Applied Phycology*, p. 429, The Hague: SPB Academic.

MAHADEVASWAMY, M. and VENKATARAMAN, L. V. (1986) Bioconversion of poultry droppings for biogas and algal production, *Agricul. Wastes*, **18**, 93.

MARQUEZ, F. J., SASAKI, K., KAKIZONO, T., NISHIO, N. and NAGAI, S. (1993) Growth characteristics of *Spirulina platensis* in mixotrophic and heterotrophic conditions, *J. Ferment. Bioengin.*, **76**, 408.

MARTINEZ, J. and BURTON, C. (1994) Traitement du lisier de porc par aération-sédimentation, *Informations Techniques du CEMAGREF*, **94**, 1.

MOHN, F. N. (1988) Harvesting of micro-algal biomass. In BOROWITZKA, M. A. and BOROWITZKA, L. J. (Eds) *Micro-algal biotechnology*, p. 395, Cambridge: Cambridge University Press.

NEWELL, P. J. (1980) The use of high-rate contact reactor for energy production and waste treatment from intensive livestock units. In VOGT, F. (Ed.) *Energy Conservation and Use of Renewable Energies in the Bio-Industries*, p. 395, New York: Pergamon Press.

OLGUIN, E. J. and VIGUERAS, J. M. (1981) Unconventional food production at the village level in a desert area of Mexico. In *Proc. 2nd World Congress of Chemical Engineering*, Montreal, Canada, 4–9 October, 1981, **1**, p. 332.

OLGUIN, E. J., HERNANDEZ, B., ARAUS, A., CAMACHO, R., GONZALEZ, R., RAMIREZ, M. E., GALICIA, S. and MERCADO, G. (1994a) Simultaneous high biomass protein production and nutrient removal using *Spirulina maxima* in sea water supplemented with anaerobic effluent, *World J. Microb. Biotechnol.*, **10**, 576.

OLGUIN, E. J, HERNANDEZ, B., ARAUS, A., CAMACHO, R., GONZALEZ, R., RAMIREZ, M. E., GALICIA, S. and MERCADO, G. (1994b) Production of *Spirulina* on sea water supplemented with anaerobic effluents from pig waste. In LALIBERTÉ, G. and DE LA NOÜE, J. (Eds) *Microalgae: From the Laboratory to the Field*, p. 145, Ste-Foy, Canada: Department of Food Science and Technology, Université Laval.

ORON, G., SHELEF, G. and LEVI, A. (1979) Growth of *Spirulina maxima* on cow-manure wastes, *Biotechnol. Bioeng.*, **21**, 2169.

OSWALD, W. J. (1988a) Micro-algae and waste-water treatment. In BOROWITZKA, M. A. and BOROWITZKA, L. J. (Eds) *Micro-algal Biotechnology,* p. 305, Cambridge: Cambridge University Press.

OSWALD, W. J. (1988b) The role of microalgae in liquid waste treatment and reclamation. In LEMBI, C. A. and WAALAND, J. R. (Eds) *Algae and Human Affairs,* p. 255, Cambridge: Cambridge University Press.

OSWALD, W. J. (1991) Introduction to advanced integrated wastewater ponding systems. *Water. Sci. Technol.,* **24**, 1.

PANO, A. and MIDDLEBROOKS, J. (1982) Ammonia nitrogen removal in facultative wastewater stabilization ponds, *Journal WPCF*, **54**, 344.

PARKER, N. C., BATES, M. C. and FEDLER, C. B. (1992) Integrated aquaculture based on *Spirulina*, livestock wastes, brine and power plant byproducts. In BLAKE, J., DONALD, J. and MAGETTE, W. (Eds) *National Livestock, Poultry and Aquaculture Waste Management*, American Society of Agricultural Engineers, Publ. 03-92, 369.

POULIOT, Y., TALBOT, P. and DE LA NOÜE, J. (1986) Biotraitement du purin de porc par production de biomasse de *Spirulina*, *Entropie*, **130/131**, 73.

PROULX, D., LESSARD, P. and DE LA NOÜE, J. (1994) Tertiary treatment of secondarily treated urban wastewater by intensive culture of *Phormidium bohneri*, *Environ. Technol.,* **15**, 449.

RICHMOND, A. (1990) Large scale microalgal culture and applications. In ROUND, F. E. and CHAPMAN, D. J. (Eds) *Progress in Phycological Research*, **7**, 269, Bristol: Biopress.

RICHMOND, A., KARG, S. and BOUSSIBA, S. (1982) Effects of bicarbonate and carbon dioxide on the competition between *Chlorella vulgaris* and *Spirulina platensis*, *Plant and Cell Physiol.,* **23**, 1411.

SAXENA, P. N., AHMAD, M. R., SHYAM, R. and AMLA, D. V. (1983) Cultivation of *Spirulina* in sewage for poultry feed, *Experientia*, **39**, 1077.

SAXENA, P. N., AHMAD, M. R., SHYAM, R., SRIVASTAVA, H. K., DOVAL, P. and SINHA, D. (1982) Effect of feeding sewage-grown *Spirulina* on yolk pigmentation of white leghorn eggs, *Avian Research*, **66**, 41.

SHAPIRO, J. (1990) Current beliefs regarding dominance by blue-greens: The case for the importance of CO_2 and pH, *Verh. Internat. Verein. Limnol.,* **24**, 38.

SHUBERT, L. E. (1988) The use of *Spirulina* (Cyanophyceae) and *Chlorella* (Chlorophyceae) as food sources for animals and humans. In ROUND, F. E. and CHAPMAN, D. J. (Eds) *Progress in Phycological Research*, p. 237, Bristol: Biopress.

SHULER, M. L. (1980) Utilization of farm wastes for food. In SHULER, M. L. (Ed.) *Utilization and Recycle of Agricultural Wastes and Residues*, p. 67, Boca Raton, Fl.: CRC Press.

SIEW-MOI, P. (1987) Agro-industrial wastewater reclamation in Peninsular Malaysia, *Arch. Hydrobiol.,* **28**, 77.

SWEETEN, J. M. (1991) Livestock and poultry waste management: a national overview. In BLAKE, J., DONALD, J. and MAGETTE, W. (Eds) *National Livestock, Poultry and Aquaculture Waste Management*, Proceedings of the National Workshop, 29–31 July 1991, ASAE Publication 03-92, 1992, 4.

TAM, N. F. Y. and WONG, Y. S. (1990) The comparison of growth and nutrient removal efficiency of *Chlorella pyrenoidosa* in settled and activated sewages, *Environmental Pollution*, **65**, 93.

TANTICHAROEN, M., BUNNAG, B. and VONSHAK, A. (1993) Cultivation of *Spirulina* using secondary treated starch wastewater, *Australasian Biotechnol.,* **3**, 223.

THOMAS, S. (1982) Algae cultivation for food and feeds. In VOGT, F. (Ed.) *Energy Conservation and Use of Renewable Energies in the Bio-Industries 2*, p. 649, New York: Pergamon Press.

VENKATARAMAN, L. V., MADHAVI DEVI, K., HAHADEVASWAMY, M. and MOHAMMED

KUNHI, A. (1982) Utilization of rural wastes for algal biomass production with *Scenedesmus acutus* and *Spirulina platensis, India Agricul. Wastes*, **4**, 117.

VONSHAK, A., GUY, R., POPLAWSKY, R. and OHAD, I. (1988) Photoinhibition and its recovery in two strains of the Cyanobacterium *Spirulina platensis, Plant Cell Physiol.*, **29**, 721.

WILDE, E. W. and BENEMANN, J. R. (1993) Bioremoval of heavy metals by the use of microalgae, *Biotech. Adv.*, **11**, 781.

WU, J. F. and POND, W. G. (1981) Amino acid composition and microbial contamination of *Spirulina maxima*, a blue-green alga, grown on the effluent of different fermented animal wastes, *Bull. Environ. Contam. Toxicol.*, **27**, 151.

YANG, P. Y. and DUERR, E. O. (1987) Bio-process of anaerobically digested pig manure for production of *Spirulina* sp., in *Summer Meeting of the American Society of Agricultural Engineers*, Baltimore, Md., June 28–July 1.

10

The Chemicals of *Spirulina*

ZVI COHEN

Introduction

Microalgae could be utilized for the production of several chemicals which are either unique to the algae or found at relatively high concentrations and command a high market value. In this respect, *Spirulina* is one of the more promising microalgae. It is especially rich, relative to other sources, in the polyunsaturated fatty acid γ-linolenic acid (GLA), and in pigments such as phycocyanin, myxoxanthophyl and zeaxanthin.

Spirulina is an outstanding candidate for outdoor cultivation, being both thermophilic and alkalophilic, with optimum growth conditions being 35–37 °C and pH of about 10; these properties facilitate the maintenance of monoalgal cultures outdoors. The high optimum growth temperature is also responsible for increased productivity. The unique chemical composition of *Spirulina* offers an additional benefit for the large-scale cultivation of this alga. This overview will discuss the chemical composition of *Spirulina* with special emphasis on chemicals of economic value.

Lipids

Typical of cyanobacteria, which are generally poor in lipids, *Spirulina* contains only 6–13 per cent lipids, half of which are fatty acids.

Monogalactosyldiacylglycerol (MGDG), sulfoquinovosyldiacylglycerol (SQDG) and phosphatidylglycerol (PG) are the major lipids, amounting to 20–25 per cent each (Petkov and Furnadzieva, 1988). Digalactosyldiacylglycerol (DGDG) contributed only 7–10 per cent, while triglycerides were a minor component (1–2 per cent). Other lipids, such as the various pigments, do not contain fatty acids. The existence of small amounts (<1 per cent) of fatty acid methyl esters was also claimed, but this result was possibly an experimental artefact.

γ-Linolenic Acid (GLA)

Properties

The rare polyunsaturated fatty acid (PUFA) γ-linolenic acid ($18:3\omega6$, GLA) is claimed to have medicinal properties. GLA lowers low-density lipoprotein in hypercholesterolemic patients (Ishikawa et al., 1989), being about 170-fold more effective than linoleic acid, the major constituent of most polyunsaturated oils (Huang et al., 1984). In addition, it has been used for the treatment of atopic eczema (Biagi et al., 1988), and for alleviating the symptoms of premenstrual syndrome (Horrobin, 1983). It is also thought to have a positive effect in heart diseases, Parkinson's disease and multiple sclerosis (Dyerberg, 1986).

Occurrence

GLA is found in higher plants – evening primrose (Wolf et al., 1983), blackcurrant (Traitler et al., 1984) and borage (Wolf et al., 1983) – as well as in cyanobacteria (Nichols and Wood, 1968) and fungi (Shimizu et al., 1988). However, large-scale cultivation of the fungi is not yet feasible, and in plant oils GLA is present either in low concentrations (8–12 per cent) or is associated with other undesirable fatty acids (e.g. $18:4$), from which separation could become very expensive on a large scale.

Nichols and Wood (1968) have shown that *Spirulina* is unique in that it contains substantial quantities of GLA. In *Spirulina*, GLA apparently plays a role normally filled by α-linolenic acid ($18:3\omega3$, ALA) in algae and higher plants. The occurrence of GLA, a precursor of arachidonic acid in animals, has led Nichols and Wood to suggest that some cyanobacteria are the origin from which both red and green algae evolved. Later works claimed that the proportion of GLA in the fatty acids of *Spirulina* is rather low (Hudson and Karis, 1974), or that ALA rather than GLA is the main polyunsaturated fatty acid (PUFA) in this alga (Kenyon and Stanier, 1970). Ciferri (1983) was the first to suggest that *Spirulina* can be used as a source of PUFAs, and especially of GLA. It is now clear, however, that *Spirulina* is the richest algal source of GLA.

Cohen et al. (1987) evaluated the fatty acid composition of 19 different *Spirulina* strains. All the tested strains except *S. subsalsa* contained the same fatty acids, and great diversity was found in the fatty acid distribution among the various strains. In keeping with earlier reports (Nichols and Wood, 1968; Hudson and Karis, 1974), the predominant fatty acids were palmitic acid ($16:0$), GLA, linoleic acid ($18:2$) and oleic acid ($18:1$) (Table 10.1). While the proportion of $16:0$ was consistent at about 44 per cent, the percentage of C_{18} fatty acids varied greatly: $18:1$ ranged from 1 to 15.5 per cent; $18:2$ between 11 and 31 per cent and GLA between 8 and 32 per cent.

Environmental and nutritional effects

Environmental conditions have a major effect on the fatty acid composition and content. Thus, manipulation of these conditions was studied in attempts to optimize GLA content and productivity.

Effect of temperature   As the cultivation temperature increased, the fatty acid content of the tested strains increased (Cohen et al., 1987), reaching the highest

Table 10.1 Fatty acid (FA) content of *Spirulina* strains grown at 35 °C

Strain	Fatty acid composition (% of total fatty acids)						Total FA (% of biomass)	GLA (% of biomass)
	16:0	16:1	18:0	18:1	18:2	GLA		
SB	44.6	4.4	0.5	6.4	17.1	27.0	5.2	1.4
Mad	47.0	0.5	0.7	9.3	10.8	31.7	4.2	1.3
Cat	47.6	2.5	1.0	8.0	15.3	25.6	5.1	1.3
Art. B	46.1	1.0	1.6	10.9	13.6	26.8	4.7	1.3
1928	47.3	2.0	1.0	2.9	18.1	28.7	4.3	1.2
L1	45.0	1.4	1.0	15.5	16.4	20.7	5.6	1.2
AR	49.1	2.2	1.0	6.4	15.7	25.6	4.3	1.1
B4	49.6	2.1	0.7	5.0	16.5	26.1	3.9	1.0
B2	47.3	3.4	0.8	5.8	20.7	20.7	3.8	1.0
G	49.2	2.9	0.9	8.0	15.7	23.3	4.0	0.9
PC	52.5	2.4	0.8	7.2	14.0	23.2	4.0	0.9
B3	52.9	2.2	1.1	7.6	13.7	22.5	4.1	0.9
Art. A	48.5	2.4	1.3	6.0	15.8	26.0	3.4	0.9
Eth	54.1	2.6	1.0	7.7	13.5	21.3	4.1	0.9
L2	50.7	1.1	0.8	7.3	14.3	25.8	3.0	0.8
Minor	46.8	1.2	1.5	12.0	18.4	20.1	3.6	0.7
2342	47.5	1.6	0.5	9.3	21.8	19.3	3.8	0.7
2340	49.3	2.2	1.2	8.6	30.7	8.0	3.2	0.3
Subsalsa	49.2	35.0	1.7	1.0	13.1	—	1.6	—

From Cohen et al., 1987, with permission.

value for most strains at 30–35 °C (Table 10.2), and the fatty acids became more saturated. The increase in saturation was partly the result of an increase in 16:0 at the expense of 16:1 and only partly due to a shift from GLA to its precursors 18:2 and 18:1. The increase in the fatty acid content (percentage of dry weight) exceeded the decrease in GLA proportion (percentage of fatty acids), and consequently there was an increase in the GLA content, which reached a maximum in most of the tested strains at 30–35 °C. This finding has practical implications in the commercial production of GLA from *Spirulina*.

Effect of light and cell concentration   The fatty acid content was found to increase with lower light intensity although the distribution was not greatly affected by the light intensity (Cohen et al., 1987).

   Increasing the cell concentration, under either batch or semi-continuous growth conditions, resulted in enhanced fatty acid, and hence increased GLA, contents (Figure 10.1). However, the effect of the increase in fatty acid content on the GLA content was partly negated by a slight decrease in the proportion of GLA (Cohen et al., 1993a), as described below. GLA was primarily concentrated in the polar lipids, especially in the galactolipids (92 per cent of total GLA), while the neutral lipids contained only very low proportions of cellular GLA (1.5 per cent). The decrease in the proportion of GLA may indicate a larger share of neutral lipids in the newly synthesized lipids. However, the increase in total lipids was not exclusively caused by an increase of neutral lipids, since the overall increase in GLA content indicates a

Table 10.2 Fatty acid (FA) content of *Spirulina* strains grown at various temperatures

Strain	Growth temperature (°C)	16:0	16:1	18:1	18:2	GLA	Total (% of biomass)	GLA (% of biomass)
Art. B	25	41.2	7.1	3.6	17.6	28.2	2.7	0.8
	30	44.5	5.1	7.2	14.5	29.7	3.4	1.0
	35	46.1	1.0	10.9	13.6	26.8	4.7	1.3
	38	53.7	2.0	8.2	13.2	21.3	3.9	0.8
G	20	39.5	9.6	1.8	16.4	32.1	2.7	0.9
	25	40.6	8.6	2.2	16.0	31.7	3.2	1.0
	30	44.6	3.1	5.1	17.9	29.0	4.0	1.1
	35	49.2	2.9	8.0	15.7	23.3	4.0	0.9
	38	47.8	0.2	9.3	17.0	25.0	3.3	0.8
Minor	25	43.2	6.9	4.8	20.5	22.2	4.8	1.1
	30	46.2	4.6	9.0	15.0	24.5	3.3	0.8
	35	46.8	1.2	12.0	18.4	20.1	3.6	0.7
	38	47.7	1.1	9.9	22.4	18.2	3.2	0.6
2340	25	43.5	7.9	4.3	21.4	22.9	2.4	0.6
	30	44.4	5.5	4.2	28.6	16.9	4.2	0.7
	35	49.3	2.2	8.6	30.7	8.0	3.2	0.3
	38	50.0	2.0	7.1	32.3	6.0	3.3	0.2

The header spans: **Fatty acid composition (% of total fatty acids)** over the columns 16:0, 16:1, 18:1, 18:2, GLA.

From Cohen et al., 1987, with permission.

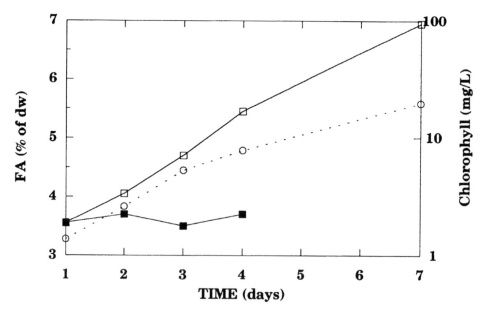

Figure 10.1 Effect of cell concentration on fatty acid content in *Spirulina* cultures. Fatty acid content was determined in batch cultures (□) and in cultures subjected to daily dilution (■). Cell density in batch cultures is expressed as chlorophyll concentration (mg L^{-1}) (O). In the daily diluted culture the chlorophyll concentration was kept at about 2 mg L^{-1}. (From Cohen et al., 1993a.)

net increase in galactolipids. At higher cell concentrations, the culture had already reached the stationary phase, and the increase in fatty acid content apparently resulted primarily from an increase in neutral lipids, hence the decrease in both the proportion and content of GLA. This effect could be related to the reduction in growth rate at higher cell densities or low light intensities, or could directly evolve from the effect of light intensity on cell biochemistry. However, reduction of the growth rate by increasing the salinity did not enhance the fatty acid content, which remained unaffected at low salinities (0.1–0.75 $g L^{-1}$) and even decreased at high salinities (1.5 $g L^{-1}$ and beyond) (Cohen et al., 1993a).

Effect of growth phase   Polar lipids are membrane components with an active role in cell function. Thus, under conditions promoting intensive growth, the proportion of polar lipids, particularly galactolipids, can be expected to reach a maximum. In the stationary phase, however, accumulation of neutral lipids may take place, perhaps as reserve material.

The major effect noted during aging from the early log phase to the stationary phase was a continuous decrease in the level of desaturation of C_{18} fatty acids (Table 10.3) (Tanticharoen et al., 1994). The proportions of GLA and 18:2 decreased, while the proportions of 18:0 and 18:1 increased. The level of 16:0, which was stable at 48–50 per cent during the early logarithmic phase, declined in the late logarithmic phase and even further so in the stationary phase, to a final level of 24 per cent. Alterations in fatty acid composition at the stationary phase can be best explained by a shift in the fatty acid flux from the production of polar

Table 10.3   Fatty acid content and distribution profile of *Spirulina* strains at various growth phases (32 °C)

Strain	Day	Growth phase[a]	Fatty acid composition (% of total fatty acids)						Total (% of biomass)	GLA (% of biomass)
			16:0	16:1	18:0	18:1	18:2	GLA		
BP	0.8	EL	48.2	0.7	1.0	5.4	19.0	24.5	2.6	0.6
BP	2	L	50.1	0.8	0.9	7.8	17.0	23.4	4.6	1.1
BP	3	L	49.5	0.9	1.1	8.9	14.6	23.6	5.1	1.2
BP	4	L	49.8	0.8	1.1	12.7	12.7	21.1	6.3	1.3
BP	6	LL	43.2	4.3	0.9	11.3	12.4	20.7	5.7	1.2
BP	14	S	23.9	2.3	3.8	30.5	10.4	20.4	5.1	1.0
P4P	1	EL	48.4	3.1	1.2	11.3	13.0	20.1	5.4	1.1
P4P	2	L	48.8	3.7	0.7	7.7	14.3	22.2	5.4	1.2
P4P	4	L	49.1	3.8	0.7	8.9	14.0	21.3	5.8	1.2
P4P	8	LL	47.2	4.7	0.7	8.9	13.2	22.6	6.4	1.4
P4P	12	S	29.9	2.6	3.0	28.5	10.7	22.7	5.3	1.1
Z6	1	EL	47.7	4.4	1.0	7.6	14.6	22.4	5.3	1.2
Z6	2	L	46.8	4.3	1.5	6.6	14.7	22.4	5.7	1.3
Z6	4	L	46.3	4.2	0.8	8.4	14.8	22.6	6.2	1.4
Z6	8	LL	47.4	4.4	0.8	7.3	14.8	22.3	6.3	1.4
Z6	12	S	31.2	3.3	2.6	22.3	12.1	25.5	5.1	1.3

[a] EL = early log phase; L = log phase; LL = late log phase; S = stationary phase.
From Tanticharoen et al., 1994, with permission.

lipids to neutral lipids (Cohen et al., 1993b). While GLA and 18:2 are primarily present in galactolipids and phospholipids, respectively, 18:0 and 18:1 are mainly found in neutral lipids. Also, in cyanobacteria, position 2 of the galactolipid glycerol backbone is predominantly occupied by 16:0 (Murata and Nishida, 1987). Therefore, a shift from production of galactolipids to neutral lipids will be manifested by an increase in 18:1 and 18:0 and a decrease in 16:0, 18:2 and GLA.

Let us look at the *Spirulina* wild-type strain BP as a typical example. The increase in fatty acid content during the growth cycle was more intense than the comparatively moderate decrease in the proportion of GLA. Consequently, the GLA content gradually increased from 0.6 per cent (of dry weight) in the early log phase to 1.2–1.3 per cent in the log phase, but was reduced to 1.0 per cent in the stationary phase (Table 10.3).

Light/dark cycles Cultures of *Spirulina* BP and derived mutants were batch cultivated under either continuous light or light/dark (12-h/12-h) cycles (Table 10.4) (Tanticharoen et al., 1994). On transfer to a light/dark cycle the cultures displayed a significant increase in C_{18} desaturation (Table 10.4). During the dark period, some of the biomass was consumed by respiration (Vonshak, 1986). The consumed biomass consisted of reserve materials such as carbohydrates and neutral lipids. A decrease in the former will result in a relative increase in the fatty acid content, while a decrease in the latter will reduce the proportion of typical triacylglycerol fatty acids such as 16:1 and 18:1, thus increasing the share of fatty acids of the major galactolipids, i.e. GLA, 16:0 and 18:2. Indeed, by cultivation of *Nannochloropsis oculata* under L/D cycles the contribution of triacylglycerols (TG) in the alga was shown to decrease at the beginning of the light period (Sukenik and Carmeli, 1990).

Table 10.4 Comparison of fatty acid composition and GLA content in various *Spirulina* mutants cultivated under continuous light or light/dark conditions

Mutant[a]	Light regime[b]	Fatty acid composition (% of total fatty acids)						Total FA (% of biomass)	GLA (% of biomass)
		16:0	16:1	18:0	18:1	18:2 ω6	18:3 ω6		
BP	CL	43.2	4.3	0.9	11.3	12.4	20.7	5.7	1.18
BP	L/D	43.8	3.0	1.1	1.2	16.6	24.7	6.1	1.62
Z5	CL	44.2	3.4	0.9	9.1	13.6	21.3	4.6	0.99
Z5	L/D	46.9	3.6	0.9	3.4	17.6	24.6	6.0	1.47
Z12	CL	47.6	3.3	1.1	10.1	14.2	22.2	4.9	1.08
Z12	L/D	45.1	3.9	0.7	2.6	19.3	27.2	5.9	1.60
Z14	CL	46.9	4.5	1.2	10.7	13.2	20.3	5.3	1.10
Z14	L/D	47.0	3.9	0.9	3.2	17.5	24.8	5.8	1.50
Z17	CL	49.3	3.4	1.0	10.0	13.3	20.3	5.1	1.03
Z17	L/D	48.5	3.6	0.7	2.3	18.3	24.0	6.2	1.49

[a] Compared cultures were of similar biomass concentration.
[b] CL = continuous light; L/D = light/dark (12-h/12-h) cycles.
From Tanticharoen et al., 1994, with permission.

It appears that cultivation under light/dark conditions, which currently is the only practical means for outdoor cultivation, is preferential to cultivation under continuous light. Moreover, it is conceivable that highest GLA contents will be obtained by harvesting at the end of the dark period.

Hirano et al. (1990) have shown that by keeping *Spirulina* cultures in the dark for a week, preferential consumption of sugars over fatty acids during the dark period resulted in a relative increase in the fatty acid content, including that of GLA, by about 50 per cent. Thus, the extent at which net production of fatty acids or GLA occurs is not clear.

Nitrogen   Piorreck et al. (1984) reported that total lipids decreased with decreasing nitrogen content in a *Spirulina* culture. Unlike eukaryotic algae, no accumulation of triglycerides was noted, and the fatty acids were found mostly in the polar fraction. Sharenkova and Klyachko-Gurvich (1975) claimed that the fatty acid composition did not change, even after 10 days in the absence of nitrogen. Lipid synthesis was reduced and eventually even stopped. It appears that since growth was arrested and no new lipids were synthesized after the third day of nitrogen starvation, the fatty acid composition could not change. A more noticeable change would probably have been observed by a gradual nitrogen depletion. Thus, varying the nitrogen concentration cannot be used as a means for manipulation of the content and composition of lipids and fatty acids in cyanobacteria.

Cultivation in the presence of ammonium chloride inhibited growth (Manabe et al., 1992). However, 8 h after the addition of ammonium chloride (25 mM), the fatty acid content increased from 5 to 7.9 per cent.

The major increase was noted in $16:0$, whose content increased from about 2.2 to 4.0 per cent. A significant increase from 1.2 to 1.7 per cent was also noted in GLA. The ammonium-chloride-induced increase of fatty acids was also observed in the dark. Thus, Manabe et al. (1992) suggested that a secondary incubation to increase the GLA content can be carried out in a smaller space, such as indoor tanks, following the primary incubation for cell proliferation.

Outdoor cultivation   The fatty acid content of cultures grown outdoors was shown to be higher than that of laboratory-cultivated cultures (Tanticharoen et al., 1994). The fatty acid content of the mutant Z19 reached 8.2 per cent as compared with 5.9 per cent under laboratory conditions. The GLA content similarly increased from 1.7 to 2.4 per cent (Table 10.5). Similar results were obtained in several other *Spirulina* mutants as well as in the wild-type strain BP (Table 10.5). Only small differences were observed in the fatty acid composition.

Optimization of production of γ-linolenic acid   The fatty acid composition and content of *Spirulina* can be manipulated so as to increase both the proportion of GLA and its content in the biomass. For example, the GLA content of BP cells maintained at a low biomass concentration was only 0.85 per cent (Table 10.6) (Tanticharoen et al., 1994). Selection of GLA-overproducing strains yielded the mutant Z19, whose GLA content under similar conditions was 1.15 per cent (see below). Cultivation of this mutant at high biomass concentration resulted in a GLA content of 1.35 per cent, which was further increased to 1.7 per cent by cultivation under light/dark cycles. Finally, by outdoor cultivation, the GLA content was enhanced to 2.4 per cent. This value is, to the best of our knowledge, the highest GLA content ever reported for any alga.

Table 10.5 Comparison of fatty acid and GLA contents under indoor and outdoor conditions in various *Spirulina* mutants

Mutant	GLA (% total fatty acids)		TFA (% dry weight)		GLA (% dry weight)	
	I[a]	O[a]	I	O	I	O
BP	22.9[b]	28.2	5.4	5.8	1.3	1.6
P4P	22.2[b]	25.8	4.8	6.1	1.1	1.6
Z1	21.9[b]	24.4	5.2	5.4	1.1	1.3
Z18	22.5[b]	25.3	4.1	5.4	0.9	1.4
Z19	28.8[c]	29.2	5.9	8.2	1.7	2.4

[a] I = indoors; O = outdoors.
[b] Continuous light, high cell concentration.
[c] Light/dark cycles.
From Tanticharoen et al., 1994, with permission.

Table 10.6 Effect of growth conditions on GLA content in a SAN-9785-resistant *Spirulina* mutant designated Z19

Strain	Growth conditions	Total fatty acids (% dry wt)	GLA (% TFA)	GLA (% dry wt)
BP (wild type)	Low cell conc.	3.6	23.6	0.85
Z19	Low cell conc.	4.2	27.4	1.15
Z19	High cell conc.	4.8	28.1	1.35
Z19	Light/dark cycle	5.9	28.8	1.70
Z19	Outdoor culture	8.2	29.2	2.40

From Tanticharoen et al., 1994, with permission

Biosynthesis

The desaturation process in cyanobacteria is unique in that all the desaturation reactions take place in lipid-bound fatty acids (Sato and Murata, 1982). In eukaryotic plants, in contrast, the first double bond is introduced into 18 : 0 in the form of 18 : 0–ACP (acyl-carrier protein), and only further desaturation is of the type of acyl-lipid desaturation to 18 : 2 and 18 : 3ω6 (Harwood, 1988). Another distinction from eukaryotic algae is that there is no desaturation of DGDG. These lipids are obtained by further galactosylation of previously desaturated MGDG molecular species.

The major galactolipid, MGDG, is obtained from glucosyl-diacylglycerol (DAG) by epimerization of glucose to galactose, rather than by replacement of glucose by galactose. PG and SQDG, however, are apparently directly produced from phosphatidic acid (PA) and DAG, respectively (Figure 10.2) (Murata and Nishida, 1987).

Selection for GLA overproducers

One way of increasing the content of specific cell components is the use of inhibitors of selected steps in biosynthetic pathways. Generally, these compounds also inhibit growth.

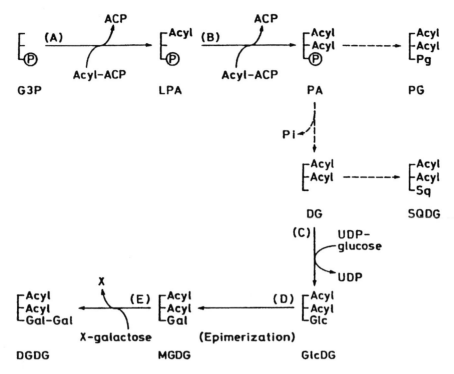

Figure 10.2 A scheme for glycerolipid biosynthesis in blue-green algae. The reactions indicated by broken arrows have not yet been experimentally demonstrated. G3P, Glycerol 3-phosphate; LPA, 1-acyl-glycerol 3-phosphate or lysophosphatidic acid; PA, phosphatidic acid; DG, 1,2-diacylglycerol; ACP, acyl-carrier protein; Acyl, fatty acid; Pg, phosphoglycerol; Sq, sulfoquinovose; Glc, glucose; Gal, galactose; X, unidentified galactose carrier. (From Murata and Nishida, 1987.)

Some overproducers of the metabolite in question were found among higher plant lines selected for resistance to the growth inhibition (Widholm, 1977; Maliga, 1980; Hibberd et al., 1980). Mutants of *Spirulina* showing elevated production of proline were obtained by selection in the presence of proline analogs (Riccardi et al., 1981). Similar results were obtained in the cyanobacterium *Nostoc* (Kumar and Tripathi, 1985) and in the alga *Nannochloris bacilaris* (Vanlerberghe and Brown, 1987).

Several herbicides of the substituted pyridazinone family have been shown to inhibit fatty acid desaturation. Of these, the substituted pyridazinone, 4-chloro-5-dimethylamino-2-phenyl-3(2H)-pyridazinone (SAN 9785) is the most effective inhibitor of $\omega 3$ desaturation (Murphy et al., 1985) and its effect on reduction of $18:3\omega 3$ levels in the glycolipids of higher plants and algae has been widely studied. The site of inhibition was proposed to be the desaturation of $18:2$ on MGDG (Willemot et al., 1982). Although SAN 9785 has a certain inhibitory effect on photosynthesis, it was shown that the effect on fatty acid desaturation is by direct inhibition of the desaturase (Wang et al., 1987). We found (Cohen, 1990) this herbicide to be an effective inhibitor for $\Delta 6$ desaturation of linoleic acid in *Spirulina*, i.e. the conversion of $18:2$ to GLA (Figure 10.3) (Cohen et al., 1993c).

We have also shown that Norflurazon, another substituted pyridazinone, which is known to inhibit carotenoid biosynthesis and chlorophyll accumulation and hence

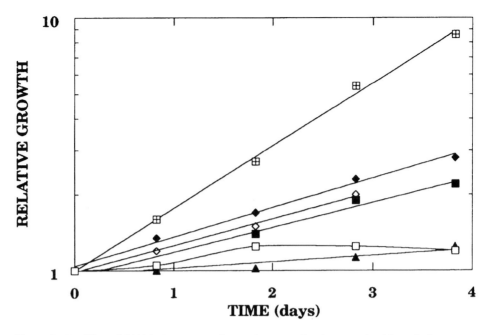

Figure 10.3 Effect of SAN 9785 on growth rate (expressed as increase in chlorophyll concentration relative to day 0) of *S. platensis* 2340 [⊞ – control (wild-type, inhibitor-free); ♦ - 0.2 mM; ■ - 0.4 mM; ▲ - 0.8 mM]. Empty symbols denote wild-type cultures freshly exposed to the herbicide. Filled symbols denote cultures resistant to the indicated concentration of SAN 9785. (From Cohen et al., 1993c.)

growth, also inhibits fatty acid desaturation, primarily Δ6 desaturation (Cohen, 1990). Its effect on the Δ6 desaturase is two orders of magnitude higher than that of SAN 9785. Over 90 per cent of the GLA in *S. platensis* is located in the galactolipids, which are important components of the photosynthetic apparatus (Cohen et al., 1993a).

Although Norflurazon is a more effective Δ6 inhibitor, we used SAN 9785 in our studies on *Spirulina*, since it also exerts a significant effect on pigment accumulation. Attempts to induce resistance to the inhibitor would require mutations in two apparently independent sites.

Maximal inhibition of growth of *S. platensis* cultures by the herbicide was obtained at a concentration of 0.8 mM SAN 9785. Even at 0.4 mM, growth was arrested after a few days and the culture collapsed thereafter. In the presence of 0.4 mM SAN 9785, the proportion of 18:3 declined from 19.6 per cent (of total fatty acids) in inhibitor-free medium to 12.4 per cent and that of 18:2 from 25.6 to 22.5 per cent, while 18:1 increased from 2.6 to 11.3 per cent (Table 10.7) (Cohen and Heimer, 1992). The fatty acid content was also affected and was reduced from 4.9 to 2.7 per cent (of dry weight), resulting in a drastic reduction of the content of GLA from 0.96 to 0.33 per cent.

The growth inhibition (via impairment of photosynthetic efficiency) probably resulted from the effect of the inhibitor on chloroplast membrane composition, since PUFAs in chloroplast galactolipids are considered to be structural components of the photosynthetic apparatus. However, we cannot rule out the possibility that growth may be adversely affected by a reduction in carbon assimilation resulting from direct inhibition of photosynthesis by the herbicide.

Table 10.7 Effect of SAN 9785 on fatty acid composition and content of *Spirulina platensis*[a]

SAN 9785 conc. (nM)	Fatty acid composition (% of total fatty acids)										Fatty acid content (% of dry wt.)	
	14:0	16:0	16:3	16:4	16:4	18:0	18:1	18:2 ω6	18:3 ω6		Total fatty acids	18:3 ω6
0	1.6	44.2	4.7	0.5	0.2	1.0	2.6	25.6	19.6		4.90	0.96
0.2	1.9	46.7	3.6	0.7	—[b]	1.9	5.0	24.3	15.9		3.90	0.62
0.4	0.9	45.0	4.0	1.0	—	3.2	11.2	22.4	12.3		2.66	0.33

[a] Fatty acid composition and content were measured following three days' exposure to the inhibitor.
[b] Trace amounts.
From Cohen and Heimer, 1992, with permission.

185

Table 10.8 Fatty acid composition of *Spirulina platensis* cell lines resistant to SAN 9785

Strain	Fatty acid composition (% of total fatty acids)						Fatty acid content (% of dry wt.)	
	16:0	16:1	18:0	18:1	18:2 ω6	18:3 ω6	Total fatty acids	18:3 ω6
Wild type	42.11 ± 0.82	5.66 ± 0.91	0.90 ± 0.05	2.06 ± 0.45	25.52 ± 0.46	21.57 ± 0.47	4.09 ± 0.42	0.88 ± 0.07
WT[a]	42.16 ± 0.93	6.29 ± 0.16	0.85 ± 0.08	2.77 ± 0.10	25.91 ± 0.47	20.21 ± 0.41	4.30 ± 0.14	0.87 ± 0.05
SRS-1[b]	43.34 ± 1.29	4.44 ± 1.89	0.70 ± 0.07	2.39 ± 0.87	24.48 ± 0.76	23.13 ± 1.18[c]	5.55 ± 0.35[c]	1.28 ± 0.13
SRS-3	42.80 ± 1.34	5.16 ± 0.22	0.84 ± 0.34	2.66 ± 1.11	24.07 ± 0.27	23.36 ± 1.21[c]	4.93 ± 0.24[c]	1.15 ± 0.09
SRS-1h	40.69 ± 1.67	4.95 ± 0.39	0.67 ± 0.04	2.58 ± 0.42	25.09 ± 0.53	23.57 ± 0.52[c]	6.07 ± 0.18[c]	1.43 ± 0.05[c]

[a] Wild type culture subjected to 0.4 mM SAN 9785 for 4 days and then transferred to inhibitor-free medium, similar to the resistant culture.
[b] Resistant cell lines selected by filament isolation from the resistant culture and cultivated on inhibitor-free medium for 11 days.
[c] Significantly different from the wild type.
Lipid transmethylation and fatty acid analyses were performed as previously reported.
Values are means ±SD ($n = 4$). A t test was used to determine significantly different values of 18:3 and TFA at $p < 0.025$.
From Cohen et al., 1992, with permission.

We have hypothesized that one means of achieving resistance to inhibitors of lipid biosynthesis would be to elicit fatty acid overproduction (Cohen et al., 1993b). Obtaining resistance to the growth inhibition caused by such compounds could provide a basis for selection of algal lines capable of GLA overproduction. The growth inhibition rendered by SAN 9785 and the specific inhibitory effects it exerted on the production of GLA led us to utilize it for the selection of resistant *Spirulina* clones, some of which were GLA overproducers.

Resistance to SAN 9785 was built up by continuous cultivation in the presence of gradually increasing concentrations of the herbicide (Cohen et al., 1993b). Thus, by this method we obtained a culture of *S. platensis* 2340 which, in the presence of 0.4 mM herbicide, had a growth rate similar to that of a wild-type culture that was freshly exposed to 0.2 mM herbicide. Eventually, the resistance was increased to withstand a concentration as high as 0.8 mM (Figure 10.3).

Relative to the wild type the resistance in *Spirulina* was associated, as predicted, with elevated levels of GLA both in the presence and in the absence of the inhibitor (Table 10.8) (Cohen et al., 1992). In inhibitor-free medium, the proportion of GLA increased from 21.6 per cent (of total fatty acids) in the wild type to 23.4 and 23.6 per cent in the resistant isolates SRS 1 and SRS 3, respectively (Table 10.8). The galactolipid fraction of SRS 1 of *S. platensis* 2340 had a high level of GLA, which reached 39.0 per cent in comparison with 33.3 per cent in the wild type (Table 10.9), while the level of 18:2 was reduced from 16.5 to 11.6 per cent (Cohen and Heimer, 1992). The two isolates had an improved fatty acid content of 6.0 per cent (of dry weight) and 4.9 per cent, respectively, compared with 4.1 per cent in the wild type. The increase in GLA content on a dry weight basis was even more marked – from 0.9 per cent in the wild type to 1.2 and 1.4 per cent in SRS 1 and SRS 3, respectively.

The resistance to SAN 9785 could be due to decreased uptake or enhanced metabolism of the inhibitor, or to reduced affinity of the target enzyme, or it could have resulted from genetic changes leading to overproduction of the target fatty acid. It appears that the latter possibility was at least a contributing factor, since elevated GLA levels were sustained for many generations after resistant cultures were shifted to herbicide-free medium.

Our data indicate that the hypothesis suggesting the use of inhibitors of fatty acid desaturation as a means for obtaining fatty acid overproduction was apparently correct. Further exploitation of this approach could be hampered by the low specificity of the herbicide. Thus, more specific inhibitors such as transition-stage analogs are being sought. We anticipate that further selection of GLA-overproducing strains in conjunction with molecular biology methodology will allow manipulation of the regulation of relevant genes, thus making algal mass production of these fatty acids feasible.

Taxonomy

Very little is known about the strains used for commercial production of *Spirulina*. Since the classification system used for *Spirulina* is essentially morphological and is mainly based on the spiral shape of the algal filaments, it is difficult to distinguish between different strains. Furthermore, the morphology of the filaments is highly dependent on growth conditions and environmental factors (Van Eykelnburg et al., 1989), and under certain conditions the filaments may even loose their spiral shape (Bai and Seshadri, 1980). The fatty acid distribution is one of the tools for

Table 10.9 Effect of SAN 9785 on fatty acid composition of the polar lipids of wild-type (BP and 2340) and SAN 9785-resistant cultures (Z19/2 and SRS-1) of *Spirulina*[a]

Lipid fraction	Strain	Medium[a]	Fatty acid composition (% of total fatty acids)									
			16:0	16:1	16:2	16:3	16:4	18:0	18:1	18:2 ω6	18:3 ω6	R[b]
Galactolipids	BP	−	41.2	4.5	—	—	2.1	1.6	4.0	9.8	36.8	11.7
	Z19/2[c]	−	41.9	5.2	—	—	—	0.8	2.7	11.1	38.3	18.5
	BP	+	44.3	1.8	—	—	0.7	2.5	19.2	9.4	22.2	1.6
	Z19/2	+	42.4	3.2	0.3	0.6	0.4	2.0	15.4	9.3	26.4	2.3
	2340	−	40.6	6.8	0.3	0.3	0.1	0.5	1.9	16.5	33.3	26.2
	SRS-1[d]	−	38.4	8.4	0.4	0.5	—	0.5	1.2	11.6	39.0	42.2
Phospholipids	BP	−	41.9	—	—	2.5	—	4.7	10.7	34.5	5.8	3.8
	Z19/2	−	41.3	3.9	0.9	0.8	1.7	2.2	7.1	34.7	7.4	5.9
	BP	+	43.6	1.0	0.7	—	0.9	3.4	28.9	18.2	3.4	0.75
	Z19/2	+	41.3	4.0	0.4	1.4	0.3	2.8	24.5	19.1	5.5	1.0

[a] (−) Inhibitor-free medium; (+) medium containing 0.4 mM SAN 9785.
[b] R: (18:3 + 18:2)/18:1.
[c] Z19/2 = culture resistant to 0.4 mM SAN 9785 derived from *S. platensis* strain BP.
[d] SRS-1 = culture resistant to 0.4 mM SAN 9785 derived from *S. platensis* strain 2340.
Data shown represent mean values of four independent samples, each analyzed in duplicate.
From Cohen et al., 1993c, with permission.

taxonomic classification of microalgae. However, its use is generally limited to the levels of classes and orders but not for genus. Kenyon and Stanier (1970) and Kenyon et al. (1972) used the fatty acid composition as a criterion for dividing cyanobacteria into subgroups and claimed the existence of two strains of *Spirulina*, one containing ALA and no GLA, and the other one devoid of both fatty acids. Based on these data, the validity of the commonly used classification system was questioned by Wood as long ago as 1974. Cyanobacterial strains were classified by Murata (Murata and Nishida, 1987; Murata et al., 1992) into four groups according to the degree of unsaturation of their polar lipids C_{18} fatty acids. In all four groups the sn-2 position of the glycerol backbone contained C_{16} fatty acids, predominantly 16:0 and 16:1. Group 3 cyanobacteria, to which *Spirulina* belongs, has $18:3\omega6$ as the major C_{18} fatty acid in MGDG and DGDG, while 18:2 is mainly a fatty acid of SQDG and PG. *S. platensis*, like other cyanobacteria, is devoid of trans $16:1\omega3$, which is found in PG of eukaryotes. The classification of cyanobacteria into four groups is in keeping with that of Kenyon et al. (1972), except that unicellular and filamentous strains may belong to each of the four groups. Recently, Pham Quoc et al. (1993) have shown that, contrary to all previous reports, *Spirulina* is capable of synthesizing eukaryotic molecular species of glycerolipids. Molecular species of MGDG from oleate-supplemented cultures consisted of only 74 per cent of the 'prokaryotic' C18/C16, and the complementary 26 per cent were C18/C18, the so-called 'eukaryotic' type of lipids. These results indicate that the specificity of the 1-monoacylglycerol-3-phosphate-acyltransferase in *Spirulina* is not as high as previously thought.

As mentioned above, we have shown that 18 strains of *Spirulina*, obtained from culture collections or isolated from their natural habitat, had the same qualitative fatty acid composition, the major fatty acids being 16:0, 16:1, 18:0, 18:1, 18:2 and GLA (Cohen et al., 1987). The proportion of GLA was quite high (Table 10.1). All these strains were cultivated on the standard alkaline *Spirulina* medium (Zarrouk's medium, pH 9.8). These findings are in agreement with those described in most other reports (Nichols and Wood, 1968; Hudson and Karis, 1974; Pelloquin et al., 1970; Paoletti et al., 1971).

We studied the fatty acid composition of six other strains, originally characterized as *Spirulina*, based on their morphological appearance and pigment composition, yet which did not grow on alkaline medium (Cohen and Vonshak, 1991). These strains were, similarly to the *Spirulina* strains previously studied, spiral shaped and filamentous with no heterocystous cells. Four fresh-water strains (Th 11, Th 21, Th 30, Sh) displayed similar characteristics in their fatty acid composition (Table 10.10) (Cohen and Vonshak, 1991). No GLA was found in these strains, although they contained high proportions of ALA. In addition, the level of palmitoleic acid (16:1) was unusually high compared with GLA-containing *Spirulina* strains and ranged from 13.8 to 32.3 per cent. Another strain, *S. subsalsa*, which was cultivated on an artificial sea water medium (Cohen et al., 1987), had a significantly different fatty acid composition. No trace of GLA was found, and the proportion of 16:1 was 35.0 per cent. In contrast, the level of 16:1 in the alkaline strains was in the range 0.5–4.4 per cent (of total fatty acids) (Table 10.10). Furthermore, since its only PUFA, linoleic acid, contained only two double bonds, this strain could not be accommodated into any of the four subgroups to which cyanobacteria are classified. The occurrence of such strains suggests that there is a fifth group of cyanobacteria, i.e. those containing PUFA with only two

Spirulina platensis (Arthrospira)

Table 10.10 Fatty acid composition of *Spirulina*-like strains cultivated in BG 11 medium

Strain	Fatty acid composition (wt % of total fatty acids)								
	14:0	16:0	16:1	16:2	16:3	18:0	18:1	18:2	ALA[a]	GLA
Th 11	1.4	41.2	13.8	—	1.7	2.6	8.9	18.1	12.3	—
Th 21	1.8	40.6	21.2	—	0.3	1.0	3.1	13.2	18.8	—
Th 30	2.5	40.4	22.3	—	0.5	1.0	3.4	13.7	16.2	—
Sh	0.6	33.1	16.8	0.2	1.2	1.4	5.8	18.7	20.5	—
LB 2179	2.2	31.2	7.6	9.5	0.8	1.7	4.9	6.9	35.3	—
N 27	2.9	47.0	18.6	19.2	2.3	0.6	1.8	5.0	—	2.6

[a] ALA = α-linolenic acid.
From Cohen and Vonshak, 1991, with permission.

Table 10.11 Classification of cyanobacteria according to fatty acid composition

Classification	18:1	18:2	18:3ω6	18:3ω3	18:3ω3
	\multicolumn{5}{c}{C₁₈ fatty acid present}				
Group 1	+	–	–	–	–
Group 2	+	+	–	–	–
Group 3	+	+	+	–	–
Group 4	+	+	–	+	–
Group 5	+	+	+	+	+

double bonds (Table 10.11). *Spirulina* strain LB 2179 had a higher proportion of ALA (35.3 per cent) and contained a significant proportion of another fatty acid (16:2), which is found only in traces, if at all, in other strains. Only one freshwater strain (N 27), contained GLA, although its level was quite low (2.6 per cent). This strain was outstanding in its high palmitoleic acid and 16:2 levels: 18.6 and 19.2 per cent, respectively.

The fatty acids of *S. subsalsa* contained about 10 per cent of a non-disclosed isomer of linolenic acid (Al-Hasan et al., 1989). Oddly, this fatty acid was concentrated mainly in DGDG (37 per cent) and in SQDG (13 per cent) with only trace levels in MGDG. The latter lipid was made of only 16:0, 16:1 and 18:1. In most algae and especially in *Spirulina*, MGDG was found to be the main depot for 18:2 and 18:3. Following dark incubation these lipids lost almost all of their linolenic acid. However, in PG, the level of linolenic acid increased from a trace level to 6.4 per cent.

It appears that several cyanobacteria, although morphologically indistinguishable from *Spirulina*, display patterns of fatty acid distribution which are different from those in *Spirulina*, both qualitatively and quantitatively. These dissimilarities appear to be significant enough to demand the exclusion of these strains from the genus *Spirulina*. Authentic *Spirulina* strains display a significant proportion of GLA, contain no ALA, and probably have a low content of 16:1 (<10 per cent) and a very low content of 16:2 (<0.1 per cent).

Following GC-MS analysis of the fatty acids of *S. platensis*, Rezanka et al. (1983) claimed the existence of rare fatty acids such as a 15:1 acid and a branched 17:0 fatty acid. Both GLA and ALA were detected. These findings are, however, at odds with almost every other report concerning the fatty acids of *Spirulina*. It is thus possible that these fatty acids, which are typical of bacteria, could have resulted from bacterial contamination.

Purification of γ-linolenic acid

At present, *Spirulina*-derived GLA cannot, on the basis of content and cost of production, compete economically with seed oil as a dietary supplement. However, a pharmaceutical commodity is required to be concentrated and free from other fatty acids. Removal of these fatty acids could be costly. Thus, sources whose GLA concentration is higher would be advantageous. *Spirulina* contains GLA mainly in the polar lipids, primarily in the galactolipids. Separation of this fraction from the lipid mixture is much easier than separation of GLA from other similar fatty acids.

Fractionation of the *Spirulina* oil by selective washes on silica gel cartridges yielded three major fractions: neutral lipids, galactolipids and polar lipids. The galactolipid fraction contained most of the GLA. Moreover, virtually all the 16:3 and most of the 18:2 were restricted to the neutral lipids and phospholipid fractions, respectively. The galactolipid fraction was converted by transmethylation into fatty acid methyl esters, which was further treated with urea in methanol. By formation of urea inclusion complexes, almost all the saturated and monounsaturated fatty acids were effectively removed. Indeed, the GLA content of the galactolipid fraction increased to 90.5 per cent (Table 10.12) (Cohen et al., 1993a).

The silica gel fractionation process achieved yet another goal: the separation of the zeaxanthin-containing xanthophylls, which were found in the neutral lipid fraction. Since very little GLA was present in the neutral lipid fraction, the fractionation enabled the practical fractionation of the oil to GLA- and zeaxanthin-rich fractions. For obtaining high-purity GLA, the galactolipid fraction could be isolated by washing with acetone. For higher yield, the last step can be replaced with a methanol wash, resulting in elution of the galactolipid and phospholipid fractions together, which contained 93 per cent of cellular GLA. Extraction of the third valuable product, phycocyanin, should, however, take precedence in order to avoid denaturization by contact with organic solvents. An outline for an all-inclusive extraction scheme of the valuable products in *Spirulina* is described in Figure 10.4.

Sulfolipids

Recently, cyanobacterial sulfolipids (SQDG) were shown to be effective against the HIV virus (EC50 0.1–1 µg/ml) and were selected by the American National Cancer Institute (NCI) for preclinical evaluation of their possible clinical usefulness against AIDS (Gustafson et al., 1989). As of now, the mechanism of the anti-HIV activity is not known. *Spirulina*, like other cyanobacteria, contains sulfolipids. If proven to be valuable, separation of sulfolipids could be performed as an extra step in downstream processing of *Spirulina* biomass. As previously mentioned, concentration of GLA

Table 10.12 GLA concentration by lipid fractionation and urea adduct formation of *Spirulina platensis* 2340 oil

Fraction	Fatty acid composition (wt % of total fatty acids)							
	14:0	16:0	16:1	16:3	18:0	18:1	18:2	18:3
Total lipid extract (TL)	1.1	39.8	7.0	0.7	1.0	2.7	25.9	21.5
Galactolipid fraction (GL)	0.2	38.4	8.5	0.4	0.5	1.6	11.6	39.0
TL after urea treatment[a]	—	0.2	0.1	1.8	—	0.2	18.0	79.7
GL after urea treatment[a]	—	0.2	0.2	1.0	—	0.1	8.0	90.5

[a] Fatty acids methyl esters resulting from transmethylation and urea treatment of the corresponding lipid.
From Cohen et al., 1993a, with permission.

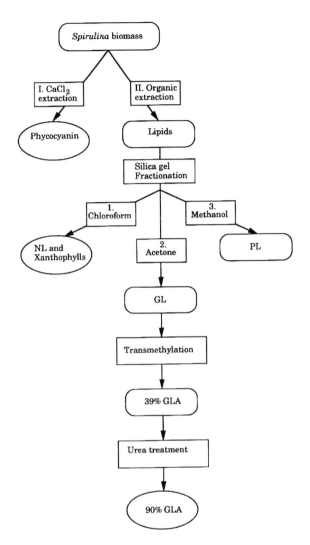

Figure 10.4 Outline for extraction and separation of valuable products from *Spirulina* biomass. (From Cohen et al., 1993a.)

can be achieved by separation of galactolipids from other lipids. Since most of the GLA resides in MGDG and DGDG, separation of the sulfolipids will result both in a more concentrated source of GLA and in the isolation of a sulfolipid fraction.

Pigments

The pigment composition of *Spirulina* is typical of cyanobacteria. The only chlorophyll present is chlorophyll *a*, its content varying from 0.8 to 1.5 per cent of dry weight (Paoletti et al., 1980). The xanthophyll content of freeze-dried *Spirulina* is considerable, reaching 6.9 $g kg^{-1}$. The other major carotenoids are myxoxanthophyll (37 per cent), a monocyclic carotenoid attached to rhamnose, β-carotene (28 per cent) and zeaxanthin (17 per cent) (Paoletti et al., 1971).

Spirulina biomass cultivated outdoors demonstrated an increase in the content of myxoxanthophyll and oscillaxanthin (Vincenzini et al., 1986). Thus, the proportion of these carotenoids which amounted to 0.006 and 15 per cent, respectively, in biomass cultivated under low light indoors increased to 6.5 and 38 per cent, respectively, outdoors. Vincenzini et al. (1986) suggested that xanthophylls, owing to their good efficiency as quenchers of the triplet sensitizers and of the singlet state of oxygen, play a major role in protecting cells against photosensitized oxidations in *Spirulina*.

Due to its pigment composition, *Spirulina* is used as a feed ingredient for pigmentation of ornamental fish, especially goldfish and fancy red carp (Miki et al., 1986). It was also used for improvement of the integumentary color of cultured fish such as tilapia and sweet melt (Miki et al., 1986) (Table 10.13). Feeding quails with a *Spirulina*-fortified diet resulted in the incorporation of xanthophylls into egg yolks, and the Roche egg yolk color score increased from 1.5 to 12.8 (Anderson et al., 1991).

In a stability study of β-carotene in *Spirulina*, Seshadri et al. (1991) found that lower drying temperature reduced decomposition of β-carotene. Shaping the final product as large flakes or granules reduced the surface area and thus exposure to oxygen and subsequent deterioration of the pigment. Addition of antioxidants and elimination of air were also found to contribute to the preservation of β-carotene.

Hydrocarbons and Sterols

The unsaponifiable fraction of the lipids contained mainly hydrocarbons and terpenic alcohols and only a low percentage of sterols (Paoletti et al., 1976). The major hydrocarbon found was n-C-17, which amounted to 64–84 per cent (of total hydrocarbons) in the different strains (Table 10.14). Nine sterols were identified by Paoletti et al. (1981) (Table 10.15), the major ones being clionasterol (33–51 per

Table 10.13 Carotenoid content and composition of *Spirulina maxima* Sml

Content (mg/100 g)	648
Composition (%)	
β-Carotene	15
β-Carotene-5,6-epoxide	+[a]
Echinenone	11–13
β-Cryptoxanthin	6–8
3'-Hydroxyechinenone	7–11
Zeaxanthin	25
Diatoxanthin	5
Canthaxanthin	5
Myxoxanthophyll	13–17
Oscillaxanthin	3–5
Total unidentified	3–4

[a] +, trace.
Adapted from Miki et al., 1986.

Table 10.14 Quantitative composition of hydrocarbons in *Spirulina platensis*

Hydrocarbon	% of total hydrocarbons	
	Strain Mao I	Strain Mao II
n-C-14	0.2	0.2
n-C-15	2.7	3.7
n-C-16	3.7	3.8
iso-C-17	3.4	3.1
n-C-17	66.9	84.0
iso-C-18	Traces	Traces
(unidentified)	0.6	0.5
(unidentified)	0.8	Traces
iso-C-19	0.4	Traces
n-C-19	0.9	1.3
Δ-C-20	0.3	Traces
iso-C-22	Traces	Traces
n-C-22	3.2	0.5
Δ-C-23	0.3	Traces
iso-C-25	0.4	—
n-C-25	1.8	0.3
n-C-26	1.3	0.2
Δ-C-27	Traces	—
(unidentified)	0.8	0.4
Squalene	0.8	0.9

Adapted from Paoletti et al., 1976

Table 10.15 Composition of the sterol fraction of the biomass of *Spirulina platensis* and *S. maxima* produced in mass culture

	S. platensis		*S. maxima*	
	Distribution (% of total sterols)	Content (mg/100 g dry wt)	Distribution (% of total sterols)	Content (mg/100 g dry wt)
Cholesterol	5.0	4.5	14.6	14.5
Brassicasterol	2.6	2.2	1.5	1.5
Δ⁵-Ergosterol	1.7	1.4	3.6	3.6
X[a]	0.6	0.5	6.5	6.5
Poriferasterol	5.5	4.6	8.0	7.9
Δ⁷-Ergosterol	3.8	3.1	7.4	7.3
Clionasterol	48.2	40.0	32.7	32.4
Condrillasterol	3.7	3.0	7.0	6.9
Δ⁵-Avenasterol	14.7	12.2	8.4	8.3
Δ⁷-Condrillasterol	8.4	7.0	8.9	8.8
Others	5.8	4.8	1.3	1.3

[a] Unidentified sterols.
Adapted from Paoletti et al., 1981.

cent of total sterols), Δ7 chondrillasterol (1.5–15 per cent) and cholesterol (4–15 per cent). However, other workers identified cholesterol and β-sitosterol as the main sterols (Martinez Nadal, 1971; Santillan, 1982).

Proteins and Amino Acids

Spirulina is especially rich in proteins, which amount to 64–74 per cent of its dry weight (Paoletti et al., 1980). The high amount of protein in *Spirulina* is unusual for a microorganism, being surpassed only by that in certain bacteria. The carbohydrate content is moderate and ranges from 12 to 20 per cent. Most of the carbohydrates are water extractable; a pH of 8.0 has been found to be most favorable for extraction, since solubility decreases at lower pH values.

Phycocyanobilins

The proteins having the highest economic potential are the biliproteins. *Spirulina* contains two biliproteins: c-phycocyanin and allophycocyanin. The protein fraction may contain up to 20 per cent of phycocyanin, a water-soluble blue pigment (Ciferri, 1983). The a and b subunits of c-phycocyanin were shown to have molecular weights of 20 500 and 23 500, respectively, while the subunits of allophycocyanin were smaller, with molecular weights of about 18 000 and 20 000. The maxima for absorption and fluorescence emission occur at 615–620 nm and 650 nm, respectively. The chromophore is phycobilin, an open tetrapyrrole. Boussiba and Richmond (1980) have shown that phycocyanin may serve in *Spirulina* as reserve material.

Phycocyanin is already being produced commercially. It was introduced in Japan as a natural coloring agent for feed and cosmetics and is produced at a rate of 600 kg/month. The main use of phycocyanin is as a food pigment. However, small quantities are used as a biochemical tracer in immunoassays, microscopy and cytometry owing to the fluorescent properties of the pigment. Phycocyanin could be utilized as a natural pigment for the food, drug, and cosmetics industries to replace the currently used synthetic pigments that are suspected of being carcinogens. Indeed, phycocyanin from *Spirulina* has already been commercialized by Dainippon Ink & Chemicals of Japan under the name of Lina blue. The product is an odorless non-toxic blue powder with a slight sweetness. When dissolved in water it is brilliant with a faint reddish fluorescence. Its color (abs. max. 618 nm) is between those of blue colors No. 1 (brilliant blue) and No. 2 (indigo carmine). It is stable from pH 4.5 to 8.0, and thermostable up to 60 °C, but exhibits poor light stability. Its uses include coloring of candy, ice cream, dairy products and soft drinks (Dainippon Ink and Chemicals Inc.). In another patent Dainippon Ink describe buffer extraction of *Spirulina* and treatment of the extract with an organic solvent, which denatures and precipitates the phycocyanin. The obtained blue pigment is used in eye shadow, eye liner and lipstick. Since it is not water soluble, it does not run when it is wet by water or sweat. It does not irritate the skin (Dainippon Ink and Chemicals Inc., 1980).

In a procedure developed by Herrera et al. (1989) the algal biomass was extracted with different aqueous salt solutions. Sodium nitrate solution brought about the highest yield of phycocyanin, while the highest purity of phycocyanin was obtained by extraction with $CaCl_2$. Treatment with activated charcoal and ultrafiltration yielded 90 per cent pure pigment. Reagent grade (50 per cent) phycocyanin was obtained by ammonium sulfate precipitation and dialysis followed by Sephadex and ion-exchange chromatography.

In one strain, *S. subsalsa*, the presence was demonstrated of the red phycobiliprotein c-phycoerythrin, which commonly occurs in red algae (Tomaselli, unpublished). When *S. subsalsa* was cultivated under a low light irradiance, the content of phycoerythrin was four times that of phycocyanins, reaching close to 10 per cent of dry weight.

Proteins for Food and Feed

The great increase in the world's human population during recent decades and man's desire for a better quality of life have made great demands on world food production, especially of protein-rich foods. Microalgae can be utilized as protein sources because of their fast growth rates and high protein contents. The cyanobacterium *Spirulina* contains up to 70 per cent proteins. It has been calculated (Bhattacharjee, 1970) that about 50 000 kg of protein per hectare could be produced annually. Also, the residue after the extraction of desirable chemicals from most algae is rich in protein and can be utilized as an animal feed.

Amino Acids

The amino acid composition has been determined for *S. maxima* (Tredici et al., 1988). The species contains 16 amino acids, eight of which are essential, including leucine (10.9 per cent of total amino acids), valine (7.5 per cent) and isoleucine (6.8 per cent). The major amino acids were glutamic acid (17.4 per cent) and aspartic acid (12.2 per cent). It appears that *Spirulina* can store amino acids. At low temperatures, growth rate is low, and the demand for amino acids is reduced (Ciferri, 1983). Excess nitrogen is metabolized into cyanophycin, a copolymer of amino acids, that is found as granules occupying up to 18 per cent of the cell volume. The polymer is made up mainly of chains of poly L-aspartic acid with arginine attached to the b-carboxyl group.

Mutant strains of *S. platensis* were induced by cultivating the species on media containing amino acid analogs (Riccardi et al., 1981). In the mutant strains that acquired resistance to analogs, it appears that the resistance was due to overproduction of the corresponding amino acids; for example, a mutant that was resistant to azetidine-2-carboxylic acid overproduced proline. Surprisingly, some of the mutants appeared to be resistant to more than one analog and to overproduce the corresponding amino acids. Cifferi (1983) suggested that by this method it would be possible to increase the proportion of some nutritionally important amino acids that the wild-type *Spirulina* does not produce in high concentrations, such as methionine. Similar to higher plants, proline overproducing mutants of *Spirulina* were able to grow in a highly saline medium, thus enabling the cultivation of *Spirulina* in this extreme environment.

Carbohydrates

Carbohydrate Composition

Carbohydrates, mainly branched polymers of glucose, amount to 15–20 per cent of the dry weight of *Spirulina* (Ciferri, 1983). According to Quillet (1975), the main sugar found is rhamnose, but later works did not confirm this finding: the most abundant sugar was found to be glucose, whose content reached 7–8 per cent (Ortega-Calvo et al., 1993), while the rhamnose content did not exceed 0.9 per cent.

Polyhydroxybutyric Acid

Polyhydroxybutyric acid (PHB) is a naturally occurring polymer found in a large number of photosynthetic and non-photosynthetic bacteria. It serves as a carbon reserve material in prokaryotes (Dawes, 1986), as does starch in eukaryotic plants. PHB is an optically active compound (Baptist and Werber, 1964), and chemical, spectral and synthetic studies (Shelton et al., 1971) support a linear head-to-tail linkage of levorotatory R-β-hydroxybutyric acid units. PHB can be used as a biodegradable thermoplastic polymer (King, 1982), its properties being similar to those of polypropylene homopolymer. Campbell et al. (1982) found that in *S. platensis*, the PHB content reached up to 6 per cent of the dry weight and was maximal at the end of the exponential phase. Addition of sodium acetate had no observable effect on the extent of PHB accumulation. However, in a later study, Vincenzini et al. (1990) found that when the alga was cultivated under photoautotrophic conditions, the PHB content did not exceed 0.3 per cent, while under mixotrophic growth conditions in the presence of acetate, PHB level amounted to about 3 per cent. The synthesis of PHB is primarily regulated by 3-ketothiolase. The latter is inhibited by high concentrations of free coenzyme A (CoA), while high concentrations of acetyl-CoA prevent the inhibition. Thus, when acetate is present in the growth medium, the intracellular acetyl-CoA concentration increases at the expense of the free CoA pool, resulting in enhancement of PHB biosynthesis. Vincenzini et al. (1990) concluded that the role of PHB in cyanobacteria is to provide cells with a mechanism for removal of excess reducing equivalents resulting from a disruption of the balanced formation of ATP and NADPH from photosynthesis (De Philippis et al., 1992).

Polysaccharides

S. platensis was shown to excrete polysaccharides into the growth medium (Filali Mouhim et al., 1993). The polysaccharides consisted of six neutral sugars, the major ones being galactose, glucose and xylose and the minor sugars being fructose, rhamnose and mannose (trace levels).

Lipopolysaccharides

The lipopolysaccharides of *Spirulina* make up 1.5 per cent of the dry weight (Tornabene et al., 1985). The lipopolysaccharides comprised 31.6 per cent

Table 10.16   Chemical composition of the lipopolysaccharides

Component	Lipopolysaccharides (% of total carbohydrates)
Glycerol	7.4
Glucose	7.5
Rhamnose	17.1
Fucose	3.3
Ribose	8.1
Xylose	4.5
3- or 4-O-Methyl hexose	8.1
Mannose	1.9
Galactose	8.2
Inositol	6.0
D-Glycero-D-mannoheptose	1.6
D-Glycero-L-mannoheptose	3.7
Unidentified	22.6
Heptose	1.4
3-Deoxy-D-manno-octulosonic acid	1.2
D-Glucosamine	2.1

Adapted from Tornabene et al., 1985

carbohydrates and 14.3 per cent fatty acids, the remaining approximately 55 per cent being unidentified. Rhamnose and galactose constituted 17 and 8 per cent, respectively, of total carbohydrates, while 23 per cent remained unidentified (Tornabene et al., 1985) (Table 10.16). The fatty acid composition was quantitatively different from that of the lipids. The major differences were noted in the very high proportion of GLA (55 per cent) and 16:1 and the presence of a hydroxy fatty acid 3-OH 14:0 (10 per cent). The proportion of 16:0 and 18:1 were significantly lower.

Other Chemicals

Vitamin B_{12}

Spirulina was found to contain a relatively high content of cyanocobalamin (Vitamin B_{12}), up to 11 mg per kg of dry weight (Ciferri, 1983). However, this vitamin is commercially produced, at a rate several orders of magnitude higher, by a mutant bacteria. Thus, it is unlikely that vitamin B_{12} would be commercially produced from *Spirulina*, even as a byproduct.

Phosphoglycerate Kinase

Phosphoglycerate kinase (PGK) may be used for the enzymatic determination of ATP. The PGK obtained from *S. platensis* is specific for ATP (Krietsch and Kuntz, 1978), unlike the commercial PGK which is not specific for ATP (Bergmeyer, 1974) and also converts GTP and ITP at slower rates. With this enzyme it is therefore possible to specifically determine ATP in tissue extracts or in mixtures of nucleotides.

Superoxide dismutase

Fridovich (1972) purified superoxide dismutase (SOD), which exists in all aerobic organisms and seems to play a major role as a health food and as a therapeutic agent for cancer and other diseases. SOD is a scavenger of free radicals, which are known to cause damage to proteins, DNA, membranes and organelles. So far, no evidence has been obtained from controlled experiments to prove its efficiency. However, if its claimed therapeutic characteristics are validated scientifically, then *Spirulina* could prove to be a useful source of this enzyme. It has been purified from *Spirulina* and could be extracted as a byproduct.

Economic Evaluation

Roughan (1989) analyzed the cost effectiveness of *Spirulina* as a source of GLA, and reached the conclusion that GLA from *Spirulina* would be 4–6 times more expensive than GLA from evening primrose. However, it appears that the samples used for that study were far from being characteristic of *Spirulina* biomass. Both their fatty acid content and GLA proportion were low, with the highest GLA content found being 0.6 per cent. In comparison, most of the *Spirulina* strains studied by Cohen et al (1987), contained higher GLA contents of up to 1.4 per cent. Several studies (Cohen et al., 1987; Cohen et al., 1993b; Hirano et al., 1990) have shown that both the composition and the content of fatty acids in *Spirulina* can be manipulated so as to significantly enhance the GLA content.

The occurrence of several other valuable chemicals in *Spirulina* further increases the economic competitiveness of GLA from this biomass. The blue pigment phycocyanin has already been commercialized in Japan as a food pigment, while the xanthophylic pigment zeaxanthin, has been shown to increase fish and shrimp pigmentation (Mori et al., 1987). *Spirulina* biomass cultivated on tapioca starch-processing effluents was recently produced in Thailand and used as a source of pigments for fish and shrimp (Tanticharoen et al., 1993). The ability to grow the biomass on the tapioca effluents resulted in a reduction of the cost of production of this biomass to 6–8 US$/kg.

The high cost of production of *Spirulina* and the low content of GLA has so far prevented it from becoming a source of the latter. However, we believe that a concerted extraction of GLA, phycocyanin and xanthophylls, concurrent with a further decrease in the cost of production, will contribute to the emergence of an economically viable process.

References

AL-HASAN, R. H., ALI, A. M. and RADWAN, S. S. (1989) Effects of light and dark incubation on the lipid and fatty acid composition of marine cyanobacteria. *J. Gen. Microbiol.*, **135**, 865.

ANDERSON, D. W., TANG, C-S. and ROSS, E. (1991) The xanthophylls of *Spirulina* and their effect, *Poultry Sci.*, **70**, 115.

BAI, J. and SESHADRI, C. V. (1980) *Arch. Für Hydrobiol.*, Beihefte Ergebnnisse Limnologie, **60**, 32.

BAPTIST, J. N. and WERBER, F.X. (1964) *SPE Trans.*, **4**, 245.

BERGMEYER, H. U. (1974) *Methoden der Enzymatischen Analyse*, Weinheim: Verlag Chemie.

BHATTACHARJEE, J. K. (1970) Micro-organisms as potential sources of food, *Adv. Appl. Microbiol.*, **13**, 139.

BIAGI, P. L., BORDONI, A., MASI, M., RICCI, G., FANELLI, C., PATRIZI, A. and CECCOLINI, E. (1988) *Drugs Expt. Clin. Res.*, **14**, 285.

BOUSSIBA, S. and RICHMOND, A. (1980) C-phycocyanin as a storage protein in the blue-green alga *Spirulina platensis*, *Arch. Microbiol.*, **125**, 143.

CAMPBELL, J. III., STEVENS, S. E. JR. and BALKWILL, D. L. (1982), Accumulation of polyhydroxybutyrate in *Spirulina platensis*, *J. Bacteriol.*, **149**, 361.

CIFERRI, O. (1983) *Spirulina*, the edible microorganism, *Microbiol. Rev.*, **47**, 551.

COHEN, Z. (1990) The production potential of eicosapentaenoic acid and arachidonic acid of the red alga *Porphyridium cruentum*, *J. Am. Oil Chem. Soc.*, **67**, 916.

COHEN, Z. and HEIMER, Y. M. (1992) Production of polyunsaturated fatty acids (EPA, ARA and GLA) by the microalgae *Porphyridium* and *Spirulina*. In KYLE, D. J. and RATLEDGE, C. (Eds) *Industrial Applications of Single Cell Oils*, pp. 243–273, Champaign, Ill.: American Oil Chemists' Society.

COHEN, Z. and VONSHAK, A. (1991) Fatty acid composition of *Spirulina* and *Spirulina*-like cyanobacteria in relation to their chemotaxonomy, *Phytochemistry*, **30**, 205.

COHEN, Z., VONSHAK, A. and RICHMOND, A. (1987) Fatty acid composition in different *Spirulina* strains under various environmental conditions, *Phytochemistry*, **26**, 2255.

COHEN, Z., DIDI, S. and HEIMER, Y. M. (1992) Overproduction of γ-linolenic and eicosapentaenoic acids by algae, *Plant Physiol.*, **98**, 569.

COHEN, Z., REUNGJITCHACHAWALI, M., SIANGDUNG, W. and TANTICHAROEN, M. (1993a) Production and partial purification of γ-linolenic acid and some pigments from *Spirulina platensis*, *J. Appl. Phycol.*, **5**, 109.

COHEN, Z., NORMAN, H. N. and HEIMER, Y. M. (1993b) Evaluation of potential use of substituted pyridazinones for inducing polyunsaturated fatty acid overproduction in algae, *Phytochemistry*, **32**, 259.

COHEN, Z., REUNGJITCHACHAWALI, M., SIANGDUNG, W., TANTICHAROEN, M. and HEIMER, Y. M. (1993c) Herbicide resistant lines of microalgae: growth and fatty acid composition, *Phytochemistry*, **34**, 973.

DAINIPPON INK and CHEMICALS INC., *Phycocyanin leaflet*.

DAINIPPON INK and CHEMICALS INC. (1980) Production of highly purified alcoholophilic phycocyanin, Japanese Patent 80 77 890.

DAWES, E. A. (1986) Microbial energy reserve compounds. In DAWES, E. A. (Ed.) *Microbial Energetics*, pp. 145–165, Glasgow: Blackie.

DE PHILIPPIS, R., SILI, C. and VINCENZINI, M. (1992) Glycogen and polyhydroxybutyrate synthesis in *Spirulina maxima*, *J. Gen. Microbiol.*, **138**, 1623.

DYERBERG, J. (1986) Linoleate-derived polyunsaturated fatty acids and prevention of atherosclerosis, *Nutr. Rev*, **44**, 125.

FILALI MOUHIM, R., CORNET, J.-F., FONTANE, T., FOURNET, B. and DUBERTRET, G. (1993) Production, isolation and preliminary characterization of the exopolysaccharide of the cyanobacterium *Spirulina platensis*, *Biotechnol. Lett.*, **15**, 567.

FRIDOVICH, I. (1972) *Accounts Chem. Res.*, **5**, 321.

GUSTAFSON, K. R., CARDELLINA, J. H., FULLER, R. W., WEISLOW, O. S., KISER, R. F., SNADER, K. M., PATERSON, G. M. L. and BOYD, M. R. (1989) AIDS-antiviral sulfolipids from cyanobacteria (blue-green algae), *J. Natl. Cancer Inst.*, **81**, 1254.

HARWOOD, J. L. (1988) Fatty acid metabolism, *Ann. Rev. Plant Physiol. Plant Mol. Biol.*, **39**, 101.

HERRERA, A., BOUSSIBA, S., NAPOLEONE. V. and HOLHBERG, A. (1989) Recovery of C-phycocyanin and γ-linolenic acid rich lipids from the cyanobacterium *Spirulina platensis*, *J. Appl. Phycol.*, **4**, 325.

HIBBERD, K. A., WALTER, T., GREEN, C. E. and GENEBACH, B. G. (1980) Selection and characterization of feed-back in sensitive tissue culture of maize plants, *Planta*, **148**, 183.

HIRANO, M., MORI, H., MURA, Y., MATSUNAGA, N., NAKAMURA, N. and MATSUNAGA, T. (1990) γ-linolenic acid production by microalgae, *Appl. Biochem. Biotechnol.*, **24**, 183.

HORROBIN, D. F. (1983) *J. Reprod. Med.*, **28**, 465.

HUANG, Y. S., MANKU, M. S. and HORROBIN, D.F. (1984) *Lipids*, **19**, 664.

HUDSON, B. J. F. and KARIS, I. G. (1974) The lipids of the alga *Spirulina*, *J. Sci. Food Agric.*, **25**, 759.

ISHIKAWA, T., FUJIYAMA, Y., IGARASHI, C., MORINO, M., FADA, N., KAGAMI, A., SAKAMOTO, T., NAGANO, M. and NAKAMURA, H. (1989) *Atherosclerosis*, **75**, 95.

KENYON, C. N. and STANIER, R. Y. (1970) Possible evolutionary significance of polyunsaturated fatty acids in blue-green algae, *Nature*, **227**, 1164.

KENYON, C. N., RIPPKA, R. and STANIER, E. Y. (1972) Fatty acid composition and physiological properties of some filamentous blue-green algae, *Arch. Microbiol.*, **83**, 216.

KING, P. (1982) Biotechnology and industrial view, *J. Chem. Technol. Biotechnol.*, **32**, 2.

KRIETSCH, W. G. and KUNTZ, G. W. K. (1978) Specific enzymatic determination of adenosine triphosphate, *Anal. Biochem.*, **90**, 829.

KUMAR, H. D. and TRIPATHI, A. K. (1985) Isolation of a hydroxyproline-secreting pigment-deficient mutant of *Nostoc* sp. by metronosazole selection, *Mircen J. Appl. Microbiol. Biotechnol.*, **1**, 269.

MALIGA, P. (1980) Isolation and characterization of mutants in plant cell culture, *Int. Rev. Cytol. (Suppl.)*, **11A**, 225.

MANABE, E., HIRANO, M., TAKANO, H., ISHIKAWA-DOI, N., SODE, K. and MATSUNAGA, T. (1992) Influence of ammonium chloride on growth and fatty acid production by *Spirulina platensis*, *Appl. Biochem. Biotechnol.*, **34**, 273.

MARTINEZ NADAL, N. G. (1971) Sterols of *Spirulina maxima*, *Phytochem.*, **10**, 2537.

MIKI, W., YAMAGUCHI, S. and KONOSU, S. (1986) Carotenoid composition of *Spirulina maxima*, *Bull. Jpn. Soc. Sci. Fish.*, **7**, 1225.

MORI, T., MURANAKA, T., MIKI, W., YAMAGUCHI, K., KONOSU, S. and WATANABE, T. (1987) Pigmentation of cultured sweet smelt fed diets supplemented with a blue-green alga *Spirulina maxima*, *Nippon Suisan Gakkaishi*, **53**, 433.

MURATA, N. and NISHIDA, I. (1987) Lipids of blue-green algae (cyanobacteria). In STUMPF, P. K. (Ed.) *The Biochemistry of Plants*, Vol. 9, p. 315, Orlando, Fl.: Academic Press.

MURATA, N., WADA, H. and GOMBOS, Z. (1992) Modes of fatty acid desaturation in cyanobacteria, *Plant Cell Physiol.*, **33**, 933.

MURPHY, J. M., HARWOOD, J. L., LEE, K. A., ROBERTO, F., STUMPF, P. K. and ST JOHN, J. B. (1985) Differential responses of a range of photosynthetic tissues to a substituted pyridazinone, Sandoz 9785. Specific effects on fatty acid desaturation, *Phytochemistry*, **24**, 1923.

NICHOLS, B. W. and WOOD, B. J. B. (1968) The occurrence and biosynthesis of gamma-linolenic acid in a blue-green alga, *Spirulina platensis*, *Lipids*, **3**, 46.

ORTEGA-CALVO, J. J., MAZUELOS, C., HERMOSIN, B. and SAIZ-JIMENEZ, C. (1993) Chemical composition of *Spirulina* and eukaryotic algae food products marketed in Spain, *J. Appl. Phycol.*, **5**, 425.

PAOLETTI, C., MATERASSI, C. and PELOSI, E. (1971) Lipid composition variation of some mutant strains of *Spirulina platensis*, *Ann. Microbiol. Enzymol.*, **21**, 65.

PAOLETTI, C., PUSHPARAJ, B., FLORENZANO, G., CAPELLA, P. and LERCKER, G. (1976) Unsaponifiable matter of green and blue-green algal lipids as a factor of biochemical differentiation of their biomasses. I: Total unsaponifiable and hydrocarbon fraction, *Lipids*, **11**, 258.

PAOLETTI, C., VINCENZINI, M., BOCCI, F. and MATERASSI, R. (1980) Composizione biochimica generale delle biomasse di *Spirulina platensis* e *S. maxima*. In MATERASSI, R. (Ed.) *Prospettive della coltura di Spirulina in Italia*, pp. 111–125, Rome: Consiglio Nazionale delle Ricerche.

PAOLETTI, C., VINCENZINI, M., BOCCI, F. and MATERASSI, R. (1981) Composizione biochimica generale delle biomasse di *Spirulina platensis* e *S. maxima*. In *Estratto da Atti del Convegno Prospettive della coltura di Spirulina in Italia*, p. 111, Firenze: Tipografia Coppini.

PELLOQUIN, A., LAL, R. and BUSSON, F. (1970) Etude comparee des lipides, de *Spirulina platensis* (Gom.) Geitler et de *Spirulina geitleri* J. de Toni, *C. R. Acad. Sci. Paris*, **271**, 932.

PETKOV, G. D. and FURNADZIEVA, S. T. (1988) Fatty acid composition of acyl lipids from *Spirulina platensis*, *Compt. Rend. Acad. Sci.*, **41**, 103.

PHAM QUOC, K. P., DUBACQ, J. P., JUSTIN, A. M., DEMANDRE, C. and MAZLIAK, P. (1993) Biosynthesis of eukaryotic lipid molecular species by the cyanobacterium *Spirulina platensis*, *Biochim. Biophys. Acta*, **1168**, 94.

PIORRECK, M., BAASCH, K-H. and POHL, P. (1984) Biomass production, total protein, chlorophylls, lipids and fatty acids of freshwater green and blue-green algae under different nitrogen regimes, *Phytochemistry*, **23**, 207.

QUILLET, M. (1975) Recherche sur les substances glucidiques élaborées par les *Spirulines*, *Ann. Nutr. Aliment.*, **29**, 553.

REZANKA, T., VOKOUN, J., SLAVÍCEK, J. and PODOJIL, M. (1983) Determination of fatty acids in algae by capillary gas chromatography–mass spectrometry, *J. Chromatog.*, **268**, 71.

RICCARDI, G., SORA, S. and CIFERRI, O. (1981) Production of amino acids by analog-resistant mutants of the cyanobacterium *Spirulina platensis*, *J. Bacteriol.*, **147**, 1002.

ROUGHAN, P. G. (1989) *Spirulina*: A source of dietary gamma-linolenic acid? *J. Sci. Food Agric.*, **47**, 85.

SANTILLAN, C. (1982) Mass production of *Spirulina*, *Experientia*, **38**, 40.

SATO, N. and MURATA, N. (1982) Lipid biosynthesis in the blue-green alga (cyano-bacterium) *Anabaena variabilis*. I: Lipid classes. *Biochim. Biophys. Acta*, **710**, 271.

SESHADRI, C. V., UMESH, B. V. and MANOHARAN, R. (1991) Beta-carotene studies in *Spirulina*, *Biores. Technol*, **38**, 111.

SHARENKOVA, H. and KLYACHKO-GURVICH, G. (1975) Changes in the composition and content of fatty acids in *Spirulina platensis* (Gom.) Geitler, grown under different conditions of nitrogen nutrition, *C. R. Acad. Agric. G. Dimitrov*, **8**, 43.

SHELTON, J. R., LENDO, J. B. and AYOSTINI, D. E. (1971) *Polymer Lett.*, **9**, 173.

SHIMIZU, S., SHINMEN, Y., KAWASHIMA, H., AKIMOTO, K. and YAMADA, J. (1988) *Biochem. Biophys. Res. Comm.*, **150**, 335.

SUKENIK, A. and CARMELI, Y. (1990) Lipid synthesis and fatty acid composition in *Nannochloropsis sp.* (Eustigmatophyceae) grown in a light-dark cycle, *J. Phycol.*, **26**, 464.

TANTICHAROEN, M., BUNNAG, B. and VONSHAK, A. (1993) Cultivation of *Spirulina* using secondary treated starch wastewater, *Austr. Biotechnol.*, **3**, 223.

TANTICHAROEN, M., REUNGJITCHACHAWALI, M., BUNNAG, B., VONKTAVEESUK, P., VONSHAK, A. and COHEN, Z. (1994) Optimization of γ-linolenic acid (GLA) production in *Spirulina platensis*, *J. Appl. Phycol.*, **6**, 295.

TORNABENE, T. G., BOURNE, T. F., RAZIUDDIN, S. and BEN-AMOTZ, A. (1985) Lipid and lipopolysaccharide constituents of cyanobacterium *Spirulina platensis* (Cyano-phyceae, Nostocales), *Mar. Ecol. Prog. Ser.*, **22**, 121.

TRAITLER, H., WINTER, H., RICHLI, U. and INGENBLEEK, Y. (1984) *Lipids*, **19**, 923.

TREDICI, R., MARGHERI, M. C., DE PHILIPPIS, R., BOCCI, F. and MATERASSI, R. (1988) Marine cyanobacteria as a potential source of biomass and chemicals, *Int. J. Solar Energy*, **6**, 235.

VAN EYKELNBURG, C., FUCHS, A. and SCHMIDT, G. H. (1989) *Anatomie Van Leewenbokek*, **45**, 369.

VANLERBERGHE, G. C. and BROWN, L. M. (1987) Proline overproduction in cells of the green alga *Nannochloris bacillaris* resistant to azetidine-2-carboxylic acid, *Plant Cell Environ.*, **10**, 251.

VINCENZINI, M., DE PHILIPPIS, R. and ENA, A. (1986) Carotenoid composition of *Spirulina platensis* and cyanospira *Rippkae* in different light conditions, *La Rivista delle Sostanze Grasse*, **LXIII**, 171.

VINCENZINI, M., SILI, C., DE PHILIPPIS, R., ENA, A. and MATERASSI, R. (1990) Occurrence of poly-β-hydroxybutyrate in *Spirulina* species, *J. Bacteriol.*, **172**, 2791.

VONSHAK, A. (1986) Laboratory techniques for the cultivation of microalgae. In RICHMOND, A. (Ed.) *Handbook of Microalgal Mass Culture*, pp. 117–145, Boca Raton, Fl.: CRC Press.

WANG, X.-M., HILDEBRAND, D. F., NORMAN, H. A., DAHMER, M. L., ST. JOHN, J. B. and COLLINS, G. B. (1987) *Phytochemistry*, **26**, 955.

WIDHOLM, J. M. (1977) Selection and characterization of amino acid analog resistant plant cell cultures, *Crop Sci.*, **17**, 597.

WILLEMOT, C., SLACK, C. R., BROWSE, J. and ROUGHAN, P. G. (1982) Effects of BASF 13-338, a substituted pyridazinone, on lipid metabolism in leaf-tissue of spinach, pea, linseed and wheat, *Plant Physiol.*, **70**, 78.

WOLF, R. B., KLEIMAN, R. and ENGLAND, R. E. (1983) *J. Am. Oil Chem. Soc.*, **60**, 1858.

WOOD, B. J. B. (1974) In STEWART, W. D. P. (Ed.), *Algal Physiology and Biochemistry*, Oxford: Blackwell Scientific.

11

Use of *Spirulina* Biomass

AVIGAD VONSHA

Introduction

The chemical composition of *Spirulina* is reviewed in detail in Chapter 10. Its high protein content and unique composition of fatty acids and vitamins are used as a basis to justify claims of its health benefits for humans. Another growing outlet for the uses of *Spirulina* products is the feed market, mainly aquaculture. This chapter does not elaborate on the health benefits of *Spirulina* for human consumption; it mainly deals with the potential of using the biomass for feed in aquaculture and poultry nutrition.

Human Consumption

As described in the Preface to this volume, the first records made of the uses of *Spirulina* were as a food and what seems to be a major source of protein supply to native tribes in South America and Africa. The re-introduction of *Spirulina* as a health food for human consumption in the late 1970s and the beginning of the 1980s was associated with many controversial claims which attribute to *Spirulina* a role of a 'magic agent' that could do almost everything, from curing specific cancer to antibiotic and antiviral activity. Since most claims were never backed up by detailed scientific and medical research, they will not be discussed in this chapter. Nevertheless, one cannot ignore the fact that more than 70 per cent of the current *Spirulina* market is for human consumption, mainly as health food. Its major claims are summarized in Table 11.1.

It is the author's current belief that although health food may present a very profitable sector of the *Spirulina* market, it is going to remain relatively a small market because of many constraints and problems in marketing, especially for producers who do not have experience or connections. Thus, any increase in volume market of *Spirulina* is going to be mainly from sales as a high-value feed additive in aquaculture and poultry nutrition rather than, what was initially predicted in the early 1970s, as a major protein supplement in human nutrition.

Table 11.1 Summary of therapeutic effects reported using *Spirulina*. More highly elaborated reports can be found in Belay et al., 1993

Effect	Experiments carried out on	Main effect	Reference
Immune system	Mouse	Experiments claiming that in mice fed with *Spirulina* a delayed type hypersensitivity can be suppressed. A *Spirulina* diet enhances the immune response by stimulating macrophage function phagocytosis and enhancing interleukin production.	Nagao et al., 1991; Hayashi et al., 1994
Reducing lipid levels in liver and serum	Rat, Human	Experiments with rats demonstrated a lower rate of increase in triglyceride level in serum and liver. When triglyceride level was induced to increase by a fructose rich diet, an addition of 5% *Spirulina* prevented the increase. In human studies, the total serum cholesterol was lowered.	Kato et al., 1984; Iwata et al., 1987, 1990; Nakaya et al., 1988; González de Rivera et al., 1993
Anticancer, antitumor effects	Hamster	Extracts of *Spirulina* or a combination of *Spirulina* and *Dunaliella* repressed or prevented the development of oral tumor.	Kato et al., 1984; Schwartz and Sklar, 1987; Schwartz et al., 1988
Antiviral activity	Cell culture, (HeLa, HEL, Vero, MDCK, MT-4)	A hot water extract from *Spirulina* (Ca-SP), which was identified as a polysaccharide containing also sulfate and calcium, was found to inhibit the replication of several enveloped viruses. It was suggested that Ca-SP selectively inhibited the penetration of virus into host cells.	Hayashi et al., 1994

Use as Feed and Feed Additives

Intensification in the growing of poultry and aquaculture is increasing the demand for specially formulated feed which will meet the high standards of the market and the need to substitute for the natural source of feed available. There are many reports on attempts to use *Spirulina* in animal feeding experiments. We will limit ours to those dealing with the addition of *Spirulina* as a specific feed component rather than the use of *Spirulina* as a total replacement for protein.

Use of Spirulina *in Poultry*

Feed additives in the poultry market, with special attention to the effect of pigments on yolk coloring, represent a major part of this fast-growing industry. A few studies on the use of *Spirulina* as a very effective agent in inducing preferred yolk color have been reported (Saxena et al., 1982, 1983).

Hens previously maintained on oxycarotenoid-free feed (control) and depleted of their carotenoids transferred dietary pigments to the egg yolk within 72 h of being fed *Spirulina*-containing diets. Yolk color intensity reached a maximum after 7 days.

The effects of various test diets on the yolk pigmentation of eggs are presented in Table 11.2. The results show *Spirulina* to be an excellent yolk pigmenter. Visual scores of the yolks produced by the birds fed *Spirulina* diets were markedly higher (13.0–14.8) than those obtained from birds fed yellow maize (4.7–8.0) and dehydrated berseem meal (4.8–5.6). Indigenous eggs indicated a higher Roche-fan score (10.8) for yolks than yellow maize and berseem meal groups. *Spirulina* gave the highest readings at all levels and produced much deeper yolk color than that produced by the highest levels of the two conventional oxycarotenoid sources.

Table 11.2 Visual yolk pigmentation scores of different treatment groups and indigenous eggs

Treatment	Raw eggs	Boiled eggs
Control		
Spirulina	1.5	1.0
6%	13.0	9.3
12%	14.2	12.9
21%	14.8	13.4
Yellow maize		
10%	4.7	2.3
20%	6.6	3.0
30%	7.4	4.3
40%	8.0	6.2
Dehydrated berseem meal		
5%	4.8	2.4
7.5%	5.6	3.9
Indigenous eggs	10.8	7.0

Adapted from Saxena et al., 1983.

Feeding with a diet containing 3 per cent *Spirulina* produced deep yellow-orange yolks that scored 13.3 on the Roche yolk color fan. Twelve per cent *Spirulina* imparted a brilliant reddish-orange color to the egg yolks with a visual score of 14.2 which increased to an average of 14.8 with algal additions up to 21 per cent, but yolk pigmentation did not increase in proportion to dietary algal concentration.

A similar trend was noticed in boiled eggs, which, however, showed lower yolk color values than raw ones. This might be due to the brightness of the vitelline membrane.

In a consumer preference study of hard boiled eggs, most of the judges preferred table egg yolk colors having a Roche-fan score between 9 and 12. These colors corresponded to feeding *Spirulina* at levels of 3–9 per cent in layer diets.

A similar study conducted in Hawaii (Ross and Dominy, 1990) compared the use of freeze-dried and extracted *Spirulina* biomass as a yolk-pigmenting agent, and its findings were further supported by a more recent study (Ross et al., 1994). It demonstrated that with increasing levels of freeze-dried *Spirulina* (Table 11.3) there was a consistent increase in yolk color. Extruded *Spirulina* also consistently increased yolk color at increasing levels. The yolk color scores of eggs from quail fed freeze-dried *Spirulina* were greater than the scores of eggs from quail fed the extruded *Spirulina* at all levels. The lower yolk color scores of the extruded series were attributed to the loss of xanthophyll during the extrusion process. The work also reports that extruding *Spirulina* with cassava meal resulted in less loss in pigmenting value of the extruded *Spirulina* than for *Spirulina* extruded with corn or barley. Whether the cassava in some way protected the xanthophylls in *Spirulina* or whether there was some physical effect of the extruding process cannot be deduced from their data. What is important is that there was no adverse effect of *Spirulina* on egg production, feed per egg, egg weight, final body weight, or mortality in any of the experiments.

It has to be pointed out that fairly effective pigmentation is achieved with relatively low concentration of *Spirulina*. More detailed study may yield an even more efficient pigmentation system. This is of importance since the addition of *Spirulina* does not necessarily mean a significant increase in the overall cost of feed to the farmer.

Use of Spirulina *in Aquaculture*

The worldwide expansion of aquaculture in the last ten years has increased the market size of inland pond-grown aquaculture products. In the artificial pond

Table 11.3 Mean Roche yolk color scores of eggs laid by Coturnix hens fed graded levels of freeze-dried or extruded *Spirulina*

Spirulina in diet	Extruded *Spirulina*	Freeze-dried *Spirulina*
(%)		
0	4.1	4.9
0.5	5.9	5.9
1.0	6.6	6.8
2.0	7.3	8.0
4.0	8.4	8.9

Adapted from Ross et al., 1994.

growing systems, a very important factor determining the economic feasibility and quality of the products is feed and the efficiency of feed utilization (Ratafia and Purinton, 1989). It is estimated that by the year 2000 the total market for feed for aquaculture will exceed the value of US$2 billion. Microalgae are a natural source of feed in the food chain of fish and many other organisms (Giwojna, 1987). Much work has been carried out in order to produce specific feed formulae for the different growing systems so as to fit not only the specific requirements of the organisms but also the various development stages that are associated with different nutritional requirements.

The most intensive studies on the use of *Spirulina* as a feed ingredient in aquaculture were performed in Japan where it was reported that already in 1989 about 100–150 tonnes of *Spirulina* were employed by Japanese fish farmers (Henson, 1990). It is also estimated that by the end of this century this market will increase by an order of magnitude, reaching a worth of US$20–30 million.

Spirulina *feed in fish nutrition*

Spirulina formulated feed increases the growth rate of many species. It improves the palatability of the feed. It was also reported (Kato, 1989) that fish fed with *Spirulina* have less abdominal fat. The feed conversion is improved as well.

It has been claimed that the fish grown on feed containing *Spirulina* are of better quality, having better flavor, firmer flesh, and brighter skin color (Hirano, 1985; Suyama, 1985; Mori, 1987).

Another very important effect relates to reduction in the rate of mortality of fingerlings or post-larval stages. An addition of 0.5 to 1 per cent of *Spirulina* in the feed has a very significant effect on growth (improvement of 17–25 per cent) and reduction of mortality (30–50 per cent), depending on species and *Spirulina* concentration. It is important to note that one of the most crucial stages in modern intensive aquaculture is the survival of the initial inoculum. Improvement in the survival rates may provide a significant improvement in the economic performance of any aquaculture farm. A report on the use of *Spirulina* in aquaculture in Japan concluded that *Spirulina* improves the cost/performance ratio of the fish feed, the largest expense in fish production.

Another set of studies (El-Sayed, 1994; Mustafa et al., 1994) relates to the use of *Spirulina* in the feed of silver sea-breams or the red sea-breams, both considered a high-value product with good market and prices. A study by Liao et al. (1990) demonstrated that an improvement in the color, taste and texture of fish fed with *Spirulina* has been observed. In the case of the sea-breams it was observed that *Spirulina* significantly increased the stromal fraction that mainly contains collagen. Collagen is one of the major constituents of intramuscular connective tissue and plays an important role in maintaining the muscle structure associated with swimming (Yoshinaka et al., 1988). It has also been observed that the carcass quality of *Spirulina*-fed fish was more acceptable than that of the control group.

Use in crustaceans

A specific application of *Spirulina*, mainly as a colorant pigmentation agent in the diet of the black tiger prawn was suggested by Liao et al. (1993). They report that incorporating 3 per cent *Spirulina* in the diet of the prawn for 14 to 28 days resulted in a marked increase in the carotenoid content in the carapace and suggested that zeaxanthin, a major component of *Spirulina* carotenoids, is rapidly converted to astaxanthin by the prawn. When the study compared *Spirulina* with the use of other carotenoid-containing feed, in every case 3 per cent *Spirulina* was the most effective pigmentation agent.

The efficiency of *Spirulina* as a feed additive to young prawns was also studied in the Fujian state fishery laboratory of China. They used two species of prawns, *Penaeus penicillatus* and *Metapenaeus sp.* in their zoea through the post-larval stage. When *P. penicillatus* in the zoea stage were fed traditional dietary feed or *Spirulina*-enriched feed, a marked increase in the survival rate was observed at the 8th day of the post-larval stage on the latter diet (Table 11.4). This increase in survival rate from 57.3 to 70 per cent actually represents an improvement of more than 22 per cent as compared with the control.

When *Metapenaeus sp.* was used and ponds were initially stocked with prawn at their mysis stage, a similar result was observed (Table 11.5). In this case the increase in the rate of survival was measured after 20 days in the post-larval stage. The increased survival from 32.5 to 47.5 per cent represents an increase of more than 46 per cent over the control.

Continuous growth in the sea-food market and requirements for specific feed can be seen in the studies on the breeding of bay scallop (*Aequipectum irradians*). Its cultivation, first introduced in 1982 in China, 5 years later reached 50 000 tonnes and was expected to double by 1995. Among feed components used in the cultivation was a mix of freshly grown microalgae such as *Phaeodactylum*, *Dictateria* and

Table 11.4   The effect of *Spirulina*-enriched feed on the survival of *P. penicillatus* in the zoea stage

Feed	Initial stock	Survival (No.)	(%)
Control	3×10^5	1.72×10^5	57.3
+0.5 ppm *Spirulina*	3×10^5	1.80×10^5	60
+1.0 ppm *Spirulina*	3×10^5	2.11×10^5	70

Survival was estimated after the 8th day of post-larval stage.
Adapted from Liao et al., 1993.

Table 11.5   The effect of *Spirulina*-enriched feed on the survival of *Metapenaeus sp.* in the mysis stage

Feed	Initial stock	Survival (No.)	(%)
Control	2×10^5	6.5×10^4	32.5
+0.5 ppm *Spirulina*	2×10^5	8.3×10^4	41.5
+1.0 ppm *Spirulina*	2×10^5	9.5×10^4	47.5

Survival was measured at the 20th day of post-larval stage.
Adapted from Liao et al., 1993.

Platymonas. These feeding systems limited the production of the bay scallops and complicated the feeding protocols.

Intensive work was carried out to test the utilization of *Spirulina* as mixed feed for abalone, scallops and penaeid shrimp (Zhou et al., 1991). It was concluded that *Spirulina* mixed feed made a good substitute for live microalgae in the cultivation of parent scallops because it proved useful for the normal development of the scallop gonads which achieved a higher fecundity and hatchery rate. Thus the use of *Spirulina* mixed feed enhanced production of the bay scallops by giving them an advantage of about one month in season. Moreover, mixed feed greatly simplified the feeding procedure, precluding the change of temperature which would normally result from pouring large quantities of five different microalgae into the production tanks. It was calculated that *Spirulina* mixed feed was cheaper than live microalgae, and some hatcheries have already adopted it in their production of bay scallop larva.

Summary

The consumption of sea food and aquaculture products in developed countries is constantly increasing. Awareness of the better nutritional quality of sea food proteins and lipids will soon make them a major source of protein in the human diet. This increased demand will cause increased production of aquaculture in artificial ponds, where feed and feeding strategies determine the product quality. The work reported here and carried out all over the world indicates that *Spirulina* can be a highly important feed component in the diet of fish and crustaceans. Its effects are very pronounced.

Much of the nutritional work is carried out locally by the hatcheries and aquaculture farms themselves and is rarely published in the literature. Once *Spirulina* gains a reputation among aquaculture farmers as an ideal feed additive, as claimed in some of the works reported, its market is going to increase by an order of magnitude. Further, there will not necessarily be a decrease in price (Ratafia and Purinton, 1989). This is why big producers of *Spirulina* may be expected to accelerate their research and development. At least one company, DIC in Japan, is already doing so.

References

BELAY, A., OTA, Y., MIYAKAWA, K. and SHIMAMATSU, H. (1993) Current knowledge on potential health benefits of *Spirulina*, *J. Appl. Phycol.*, **5**, 235.

EL-SAYED, A.-F. M. (1994) Evaluation of soybean meal, *Spirulina* meal and chicken offal meal as protein sources for silver seabream (*Rhabdosargus sarba*) fingerlings, *Aquaculture*, **127**, 169.

GIWOJNA, P. (1987) Arrow crabs: Housebreaking the narrow snouted bristle horn, *Freshwater and Marine Aquarium*, **10**, 96.

GONZÁLEZ DE RIVERA, C., MIRANDA-ZAMORA, R., DÍAZ-ZAGOYA, J. C. and JUÁREZ-OROPEZA, M. A. (1993) Preventative effect of *Spirulina maxima* on the fatty liver induced by a fructose-rich diet in rat, a preliminary report, *Life Sciences*, **53**, 57.

HAYASHI, O., KATOH, T. and OKUWAKI, Y. (1994) Enhancement of antibody production in mice by dietary *Spirulina platensis*, *J. Nutr. Sci. Vitaminol.*, **40**, 431.

HENSON, R. H. (1990) *Spirulina* algae improves Japanese fish feeds, *Aquaculture Magazine*, November/December, 38.

HIRANO, T. (1985) Effect of dietary micro-algae on the quality of cultured *Ayu plecoglossus-altiverlis*, *J. Tokyo Univ. Fish.*, **72**, 21.

IWATA, K., INAYAMA, T. and KATO, T. (1987) Effects of *Spirulina platensis* on fructose-induced hyperlipidemia in rats, *J. Jap. Soc. Nutr. Food Sci.*, **40**, 463.

IWATA, K., INAYAMA, T. and KATO, T. (1990) Effects of *Spirulina platensis* on plasma lipoprotein lipase activity in fructose-induced hyperlipidemic rats, *J. Nutr. Sci. Vitaminol.*, **36**, 165.

KATO, T. (1989) Cherry salmon fed *Spirulina*, Dainippon Ink and Chemicals, Inc., July, 1989, Tokyo, Japan, unpublished paper.

KATO, T., TAKEMOTO, K., KATAYAMA, H. and KUWABARA, Y, (1984) Effects of *Spirulina* (*Spirulina platensis*) on dietary hypercholesterolemia in rats, *J. Jap. Soc. Nutr. Food Sci.*, **37**, 323.

LIAO, W., TAKEUCHI, T., WATANABE, T. and YAMAGUCHI, K. (1990) Effect of dietary *Spirulina* supplementation on extractive nitrogenous constituents and sensory test of cultured striped jack flesh, *J. Tokyo Univ. Fish.*, **77**, 241.

LIAO, W. L., NUR-E-BORHAN, S. A., OKADA, S., MATSUI, T. and YAMAGUCHI, K. (1993) Pigmentation of cultured Black Tiger Prawn by feeding with a *Spirulina*-supplemented diet, *Nippon Suisan Gakkaishi*, **59**, 165.

MORI, T. (1987) Pigmentation of cultured sweet-smelt fed diets supplemented with a blue-green alga *Spirulina maxima*, *Bull. Jap. Soc. Fisheries Sci.*, **53**, 133.

MUSTAFA, M. G., UMINO, T. and NAKAGAWA, H. (1994) The effect of *Spirulina* feeding on muscle protein deposition in red sea bream, *Pagrus major*, *J. Appl. Ichthyol.*, **10**, 141.

NAGAO, K., TAKAI, Y. and ONO, M. (1991) Exercises of growing mice, and the effect of the intake of *Spirulina platensis* upon the hapten-specific immune response, *Sci. Phys. Power*, **40**, 187.

NAKAYA, N., HONMA, Y. and GOTO, Y. (1988) Cholesterol lowering effect of *Spirulina*, *Nutr. Rep. Int.*, **37**, 1329.

RATAFIA, M. and PURINTON, T. (1989) Emerging aquculture market, *Aquaculture Magazine*, **6-7**, 32.

ROSS, E. and DOMINY, W. (1990) The nutritional value of dehydrated, blue-green algae (*Spirulina platensis*) for poultry, *Poultry Sci.*, **69**, 794.

ROSS, E., PUAPONG, D. P., CEPEDA, F. P. and PATTERSON, P. H. (1994) Comparison of freeze-dried and extruded *Spirulina platensis* as yolk pigmenting agents, *Poultry Sci.*, **73**, 1282.

SAXENA, P. N., AHMAD, M. R., SHYAM, R., SRIVASTAVA, H. K., DOVAL, P. and SINHA, D. (1982) Effect of feeding sewage-grown *Spirulina* on yolk pigmentation of White Leghorn eggs, *Avian Research*, **66**, 41.

SAXENA, P. N., AHMAD, M. R., SHYAM, R. and AMLA, D. V. (1983) Cultivation of *Spirulina* in sewage for poultry feed, *Experientia*, **39**, 1077.

SCHWARTZ, J. L. and SKLAR, G. (1987) Regression of experimental hamster cancer by beta carotene and algae extracts, *J. Oral Maxillofac. Surg.*, **45**, 510.

SCHWARTZ, J. L., SKLAR, G., REID, S. and TRICKLER, D. (1988) Prevention of experimental oral cancer by extracts of *Spirulina-Dunaliella* algae, *Nutr. Cancer*, **11**, 127.

SUYAMA, M. (1985) Odor of *Ayu plecoglossus-altiverlis* and its volatile components, *Bull. Jap. Soc. Fisheries Sci.*, **51**, 286.

YOSHINAKA, R., SATO, K., ANBE, H., SATO, M. and SHIMIZU, Y. (1988) Distribution of collagen in body muscle of fish with different swimming modes, *Comp. Biochem. Physiol.*, **89**, 147.

ZHOU, B., LIU, W., QU, W. and TSENG, C. K. (1991) Application of *Spirulina* mixed feed in the breeding of Bay Scallop, *Bioresource Technol.*, **38**, 229.

Appendices

AVIGAD VONSHAK

Introduction

The following appendices are a result of many requests from people interested in *Spirulina* but with a limited access to the main scientific literature or to established libraries. The appendices are not intended to provide all the available information but rather answer some of the most frequent questions. They are based on information gathered in the editor's lab by his colleagues and himself. Recently, Earthrise Farms established a WWW site on the internet containing some interesting information: http:/www.earthrise.com/spirulina/

Appendix I – Chemical Analysis Procedures

Chlorophyll Determination

Sample of 5 ml algal suspension is either centrifuged for 5 min at 2000 *g* (3500 rpm) or filtered through a Whatman GF/C filter 25 mm (diameter).

The pellet is kept (supernatant is discarded), or the filter with the sample on it is resuspended in 5 ml methanol (absolute) and ground in a glass tissue homogenizer. Samples are incubated in a water bath, at 70 °C for 2 min. Then the sample is centrifuged, and the clear supernatant is used for measurement.

Optical density is measured at 665 nm. You may keep the pellet for protein determination.

Calculations: Optical density at 665 nm X Factor (derived from the absorption coefficient) = Chlorophyll concentration in mg ml^{-1} or mg l^{-1}

The factor for *Spirulina* is 13.9.

Dry Weight Determination

Sample containing 25–50 ml algal suspension is filtered through a Whatman GF/C filter 47 mm (diameter) which was dried in an oven for 24 h or overnight at 70 °C or 2 h at 105 °C and weighed prior to the filtration.

While the sample is being filtered it is washed with 20 ml acidified water (pH 4) in order to remove/wash the algae from insoluable salts.

The filter is put in a glass petri dish in the oven under the above conditions.

After cooling the filter in a desiccator (20 min) it is weighed again.

Protein Determination

A sample of 1–5 ml of culture is centrifuged or filtered. The supernatant is discarded and the chlorophyll is removed by methanol extraction as previously described. The pellet is dried by blowing a gentle stream of air or N_2. To the pellet, 2 ml of 0.5 N NaOH are added, mixed well and incubated for 20 min at 100 °C (cover tubes to avoid evaporation).

Centrifuge and keep the supernatant, add another volume of hot 0.5 N NaOH (70 °C), mix well, centrifuge and combine the supernatant.

For the color reaction use 0.1–0.5 ml from the supernatant and add 0.5 N NaOH to a final volume of 1 ml.

Use BSA as a standard in the range 50–200 mg.

Reagents for the color reaction

Prepare:

- Reagent A: 2% Na_2CO_3;
- Reagent B: 0.5% $CuSO_4 \times 5H_2O$;
- Reagent C: 1% Na-tartarate;

- Reagent D: Mix together 48 ml of R.(A) + 1 ml R.(B) + 1 ml R.(C), (prepare this mix fresh each time). Add 4 ml of R.(D) to the 1 ml sample, mix well, and wait 10 min. Then add 1 ml of the Folin-Ciocalteus reagent diluted with water 1:1. Mix well and wait 30 min. Read the absorbance at 660 nm.

DNA extraction from Spirulina platensis

500 ml of 3–4 days old culture of *Spirulina platensis* ($OD_{600} = 1$) is used for DNA extraction. Filaments are harvested by centrifugation, washed once with sterile distilled water and suspended in 2 ml lysis buffer (25 per cent sucrose, 10 mM EDTA, 50 mM Tris.HCl, pH 7.5). The suspension is frozen at $-70\,°C$ for 30 min and thawed at $37\,°C$. Lysozyme is added to a final concentration of 1 mg ml^{-1} and the suspension is incubated at $37\,°C$ for 30 min. SDS (final concentration 10 mg ml^{-1} and Proteinase K (final concentration 0.1 mg ml^{-1}) are added and the mix is incubated overnight in a water bath set at $55\,°C$. An equal volume of phenol-chloroform-isoamylalcohol is added and mixed gently for 10–15 min. The aqueous phase is collected by centrifugation and RNAse treated (final concentration of DNAse free RNAse 50 mg ml^{-1}) at $37\,°C$ for 1 h. This is followed by extraction with phenol-chloroform-isomyl alcohol and chloroform-isoamyl alcohol. Sodium acetate is added to a final concentration of 0.3 M. The DNA is collected by ethanol precipitation with two volumes of ice-cold ethanol. The pellet is rinsed with ice-cold 70 per cent and 100 per cent ethanol, air-dried, dissolved in TE (10 mM Tris.HCl, pH 8., 1.0 mM EDTA) and stored at $-20\,°C$.

Appendix II – Where to Get *Spirulina* Strains

Culture Collections

Inst. Pasteur
Service Financier et Comptable
28 Rue du Dr Roux
75724 Paris Cedex 15, France
Tel. 33-1-4568-8000
Fax. 33-1-4306-9835

Culture Collection of Algae
Department of Botany
The University of Texas at Austin
Austin TX 78713-7640, USA
Tel. 1-512-471-4019
Fax. 1-512-471-3878
E-mail. Jeff.N.Judy@mail.utexas.edu

Culture Collection of Autotrophic Organisms
Czech Academy of Sciences
Institute of Botany
Dukelska 145
CS 379 82, Trebon, Czech Republic

American Type Culture Collection
12301 Parklawn Drive
Rockville, Maryland 20852, USA
Tel. 1-800-638-6597
Fax. 1-301-231-5826

Some laboratories have their own collections, not always available for distribution:

Dr Luisa Tomaselli
Centro di Studio dei Microrganismi Autotrofi
del CNR e Instituto di Microbiologia Agraria
e Tecnica dell'Universita di Firenze
Piazzale delle Cascine
27-50144 Firenze, Italy
Tel. 39-55-36050
Fax. 39-55-330431

Drs N. Jeeji Bai and C.V. Seshadri
Shri AMM Murugappa Chettiar Research Center (MCRC)
Photosynthesis and Energy Division
Tharamani, Madras 600 113, India
Tel. 91-44-2350937, 2350369
Fax. 91-44-510378, 415856

Dr Avigad Vonshak
Micro-Algal Biotechnology Laboratory
Jacob Blaustein Institute for Desert Research
Ben-Gurion University of the Negev
Sede-Boker Campus 84990, Israel
Tel. 972-7-6565825
Fax. 972-7-6570198

Appendix III – Growth Media and Conditions for *Spirulina*

Macroelements

	$g L^{-1}$
NaCl	1.0
$CaCl_2$	0.04
$NaNO_3$	2.5
$FeSO_4 \times 7H_2O$	0.01
EDTA (Na)	0.08
K_2SO_4	1.0
$Mg\ SO_4 \times 7H_2O$	0.2
$NaHCO_3$	16.8
K_2HPO_4	0.5

+ microelement solutions
A_5 and B_6, as listed below

The various salts of the macroelements should be introduced into the solution in the written order. Phosphate should always be added last.

Microelements

A_5

	$g L^{-1}$
$Zn\ SO_4 \times 7H_2O$	0.222
$CuSO_4 \times 5H_2O$	0.079
MoO_3	0.015
H_3BO_3	2.86
$MnCl_2 \times 4H_2O$	1.81

B_6

	$g L^{-1}$
NH_4VO_3	229.6×10^{-4}
$K_2Cr_2(SO_4)_4 \times 24H_2O$	960.0×10^{-4}
$NiSO_4 \times 7H_2O$	478.5×10^{-4}
$Na_2WO_4 \times 2H_2O$	179.4×10^{-4}
$Co(NO_3)_2 \times 6H_2O$	439.8×10^{-4}
$Ti_2(SO_4)_3$	400.0×10^{-4}

Details of preparation

The freshly prepared solution should have a pH in the range 8.7 to 9.3, after sterilization.

Because of the poor solubility of NH_4VO_3, B_6 solution tends to be turbid. Thus this solution should be well stirred before usage.

Solutions A_5 and B_6 should be kept refrigerated, replacing them after 2 months.

Growth conditions

Growth temperature: 35 °C.

Spirulina may be easily photoinhibited, thus one should make sure to start the culture in dim light, i.e. 15 μE, and gradually increase irradiance.

Once the culture begins to grow, pH should be kept at about 9.8 by bubbling CO_2 ca. 1–2 per cent in air. When grown in tubular (3 cm diameter) vessels, the initial chlorophyll concentration should be 1–2 mg ml^{-1}.

A growth rate curve should be followed, and the best cell density in which to maintain the culture is at ca. $1/2\,\mu max$.

Maintenance

Cultures can be maintained on solidified medium (1.2–1.5 per cent agar). If kept at low light of 10–20 μmol $m^{-2}\,s^{-1}$ and 20 °C, cells will be viable for more than 6 months if not heavily contaminated by bacteria.

Appendix IV – Some Basic and Popular Reading on *Spirulina*

BEASLEY, S., *Spirulina Cook Book*, University of the Trees Press, Boulder Creek, Calif. (1981).

BONNIN, G., *Spirulina Production Engineering Handbook* (Cyanobacterial cultivation, harvesting and drying), B.E.C.C.M.A., France (1992).

CIFERRI, O., *Spirulina, the Edible Organism, Microbiological Reviews* (1983) 572.

FOX, R., *Spirulina* – the alga that can end malnutrition, *The Futurist*, Feb. (1985).

HENRIKSON, R., *Earth Food Spirulina*, Ronore Enterprises, Inc., Calif.

HILLS, C., *The Secrets of Spirulina*, University of the Trees Press, Boulder Creek, Calif. (1980).

JASSBY, A., *Spirulina*: a model for microorganisms as human food, in *Algae and Human Affairs* (Lemby, Waaland, Eds), Cambridge University Press, Cambridge (1988) p. 171.

NAKAMURA, H., *Spirulina: Food For a Hungry World, A Pioneer's Story in Aquaculture*, University of the Trees Press, Boulder Creek, Calif. (1982).

RICHMOND, A., *Spirulina*, in *Micro-Algal Biotechnology* (Borowitzka, L. and Borowizka, M., Eds), Cambridge University Press, Cambridge (1988).

SESHADRI, C.V. and Bai, N.J., *Spirulina, ETTA National Symposium*, MCRC, Madras, India (1992).

SWITZER, L., *Spirulina* – the Whole Food Revolution, Bantam Books, New York (1982).

VONSHAK, A. and RICHMOND, A., Mass production of the blue-green algae *Spirulina*: an overview, *Biomass*, **15**, 233 (1988).

Appendix V – The Main Commercial Producers of *Spirulina*

The information is based on data provided by the companies.

In the last few years, more production sites for *Spirulina* are reported to have started commercial production, in China, India and Vietnam.

1

Name of company	Ballarpur Industries Ltd
Mailing address	Spirulina Farm
	P.O. Box No. 07, KIADB Industrial Area
	Nanjangud - 571 301, Mysore District *India*
	(Tel. 8221-26775, 26763 ; Fax. 8221-268872)

Contact persons for further information or sales inquiries:
Mr. A. Rangarajan; Mr. B.A. Shanbhag.
The Thapar Group is one of the top corporate conglomerates in India today. Ballarpur Industries Limited (BILT) is the torch-bearer of the Thapar Group. The *Spirulina* production site near Mysore was started a few years ago and was initially based on collaborative work with Dr L.V. Venkataraman,
Dept of Plant and Cell Biotechnology, Central Food Technological Research Institute, Mysore - 570 013 *India.*

Production information
Intensive ponds, total size: 54 000 m^2.

Product information
The main product is *Spirulina* powder, 95 per cent and pills 5 per cent.
From April 94 to March 95: the total biomass produced was 25 tonnes.
It is expected that for the period April 95 to March 96 the total production will be in the range of 85 tonnes. The product is mainly sold for human consumption.

2

Name of company	Cyanotech Corporation
Mailing address	73-4460 Queen Kaahumanu, # 102
	Kailua-Kona, HI 96740, *USA*
	(Tel. 808-326-1353; Fax. 808-329-3597)

Contact persons for further information or sales inquiries:
Mr Kelly Moorhead.
Cyanotech, originally started as a *Dunaliella* production facility, is now mainly involved in the production of *Spirulina* using sea water. The Company's current products include *Spirulina Pacifica*™, a high-protein beta-carotene rich nutritional product; and phycobiliproteins, fluorescent pigments used in immunological diagnostics. Nutrex, Inc., the Company's wholly-owned subsidiary, produces and markets a line of *Spirulina*-based nutritional products for the retail market.

Production information
Intensive ponds, total area of 100 000 m^2.

Product information
The main product is *Spirulina* powder, 70 per cent and pills 30 per cent.
During 1995 the total production was 250 tonnes.
It is estimated that the 1996 production will be in the range of 300 tonnes.
The product is mainly sold for human consumption.

3

Name of company	Earthrise Farms
Mailing address	P.O. Box 270
	Calipatria, CA 92233, *USA*
	(Tel. 619-348-5027; Fax. 619-348-2895)

Contact persons for further information or sales inquiries:

Information:	Mr Yoshimichi Ota, President;
Sales Inquiries:	Earthrise Company, P.O. Box 60, Petaluma,
	CA 94953-0060, USA, Tel. 707-778-9078; Fax. 707-778-9028.

This was the first large commercial producer of *Spirulina* in the USA. The trademark of Earthrise is now registered in more than 30 countries and well accepted as the symbol for quality products in the world market place.
For the entire activities of Earthrise group, browse Internet World Wide Web Site: http:/www.earthrise.com/Spirulina/.

Production information
Intensive ponds, total area 150 000 m^2.

Product information
Products include powder, pills and formulated food; no breakdown was given.
The total production in 1995 was approximately 360 tonnes.
It is estimated that the 1996 production will be in the range of 400 tonnes.
Products are sold for human consumption and animal feed.

4

Name of company	Myanma Microalga Biotechnology Project
Mailing address	Myanma Pharmaceutical Industries,
	192 Kaba Aye Pagoda Road, Yangon, Myanmar
	(Fax. 95-1-56722)

Contact persons for further information or sales inquiries:
Managing Director, Myanma Pharmaceutical Industries.
This is the largest production site based on natural blooms. The lake Twin Taung is located in a volcanic crater with natural blooms of *Spirulina*.

Production information
Mainly native ponds with a total area of about 130 000 m^2.
The product, mainly pills, is marketed locally.
It is reported that in 1995 the total production was 32 tonnes, and it is estimated that in 1996 it will be in the range of 40 tonnes.

5

Name of company	Siam Algae Co., Ltd
	A subsidiary of Dainippon Ink & Chemicals Inc.
Corporate office:	21st Fl., Serm-mit Tower, 159 Sukhumvit 21 Rd.,
	Klongtoey, Bangkok 10110, Thailand
	Tel. 2-260-6644; Fax. 2-260-6647
Farm office	No. 19, Moo 4, Tambol Bangsaotong, Amphoe
	Bangplee, Samut-Prankton, Thailand
	Tel. 313-1643; Fax. 313-1645

Production information

Dried *Spirulina* powder, and some tablets.
Production mode:

	Intensive ponds (concrete made)	Native ponds	Total
Total Pond area	28 000 m^2	16 000 m^2	44 000 m^2
No. of ponds	2 000 m^2 × 13 p 1 000 m^2 × 2 p	3 000–3 600 × 5 p	

Product information

	Productivity (tonne/year)		
	For human consumption	Feed market	Total
1995 (proj.)	95	30	125
1996 (proj.)	100	30	130

Market: Japan and International markets except North America.
DIC is marketing *Spirulina* in the form of powder, tablets (main product), formulated foods, formulated feed, extracted blue color (phycocyanin product – Lina Blue).

6

Name of company	Wuhan Micro-alga Biotechnology Company
Mailing address	Wuhan City, Mosan, 430074 (in Wuhan Institute of Botany, Academia Sinica), P.R. China Tel. 86-27-7802680, 7877080 Fax. 86-27-7877080

Contact persons for further information or sales inquiries:
Prof. Hu Hong-jun and Mr Hu You-ming.
The Wuhan Micro Alga Biotechnology company, founded in 1991, is a high- and new-technology enterprise under the administration of the Wuhan Donghu High and New Technology Development Zone, primarily engaged in development and production of microalgae and other biotechnological products.
This company took the lead in building a high-purity *Spirulina* culture based on a geothermic temperature-regulated greenhouse which is the largest in China.
The *Spirulina* project was jointly developed by this company and the Wuhan No. 4 Pharmaceutical Factory.

Production information

The culture ponds are in the greenhouse. The total size of culture ponds covers an area of 2 hectares.

Product information

Powder, pills, capsules, formulated feed health food.
Formulated feed for shrimps, bay scallop etc.

- 20% powder, 65% for pills and capsules, 15% for feed and extraction.
- 80% for food (powder, pills and capsules), 17% for feed, 8% for extraction.

Spirulina platensis (Arthrospira)

- Total production in 1995: 25 tonnes
- Expected production in 1996: 40 tonnes
- 75% for health food, 15% for feed, and 10% for extraction.

7

Name of company	Neotech Food Co., Ltd
Mailing address	111/1 Moo 4, Cholpratan Road, Berkprai Banpong, Rajburi 70110, Thailand (Tel. 66-32-210620; Fax. 66-32-210621)

Contact persons for further information or sales inquiries:
Mr Chaivuth Suksmith; Ms Pattra Honglumphong.

Production information
Intensive ponds, total size of production area 50 000 m^2.

Product information
Spirulina powder.
In 1995 30 tonnes were produced; planning to produce 40 tonnes in 1996.
The main product is 30 per cent for human consumption and 70 per cent for animal feed.

8

Name of company	Nan Pao Resins Chemical Co., Ltd
Mailing address	491 Chungshan Rd., Shee Kang Shiang Tainan, Taiwan, R.O.C. (Tel. 886-7952621; Fax. 886-6-7952184)

Contact persons for further information or sales inquiries:
Mr K.F. Hwu.

Production information
The production mode used is intensive ponds, total size of production area 50 000 m^2.

Product information
Producing *Spirulina* powder and pills for human consumption.
In 1995 70 tonnes were produced; planning to produce 80 tonnes in 1996.
Final plans are to reach production of 150 tonnes by the year 2000.

Appendix VI – Suggested Parameters for Evaluation of *Spirulina* Quantity

	Recommended content	Comments
1. Heavy metal content		
Mercury, Cadmium	<0.1 mg l^{-1}	
Lead, Arsenic	<2.0 mg l^{-1}	
2. Microbial test		
Coliforms	Negative	
Salmonella	Negative	
Shigella	Negative	
Staphylococcus	Negative	
Mold	<100 g^{-1}	
3. Chemical composition		
Protein	50–60%	1
Phycocyanin	10–15%	
Chlorophyll *a*	1–1.5%	2
Other chlorophylls	Negative	3
Gamma-linolenic acid (GLA)	21%	4
Pheophorbides	<1.2 mg g^{-1}	
Ash	<10%	5
Moisture	<7%	
4. Filth		6
Insect fragments	<150 per 50 g	
Rodent hairs	0.5 per 50 g	
5. General		
Appearance		
– Powder	Uniform, no flakes	
– Pills	Smooth surface, no binders, no coating	7
Color	Green to dark green, highly depends on strain, no brown color	
Taste	Mild	8

Comments

1. Many values have been published on the protein content of *Spirulina* biomass. It has to be realized that figures in the range 70–80 per cent are misleading, since they are most likely a result of determining the protein content by a chemical determination of the nitrogen content and using an overestimating factor. We would like to suggest that any chemical analysis will include the procedure used for the determination.

2. Chlorophyll can be a good indicator for the 'quality' of the culture. A chlorophyll value below 1 per cent on a dry weight basis will usually indicate high zooplankton contamination or poor management of product processing.

3. Existence of chlorophyll types like *b* or *c* will be an indicator of contamination by other algae.

4. Although not medically proven, many of the beneficial effects claimed after using *Spirulina* are a result of the unique fatty acid composition. 'Good' *Spirulina*

225

strains and a good processing procedure should yield biomass with at least 1 per cent of GLA.

5. Most of the ash is a result of the chemicals used for cultivation and water quality. High ash content is a result of poor washing in the processing step.

6. These are relatively strict standards to be met as required by the FDA. A good prefilter before the harvest will help. *Spirulina* grown in tubular reactors will most likely have lower values.

7. The appearance of the product can also be a good indicator of the processing and quality control applied. Some producers add artificial binders which result in white or dark dots on the surface of the pill. The binders may affect the digestibility of the product.

8. Bitter or salty taste may indicate not enough washing or addition of preservatives.

Index